Drone University

By John M. Glover

Copyright and Disclaimer

Drone University

Copyright © 2014 by John M. Glover

Email Address: info@kiloOhm.com

ISBN-13:
978-0692316030 (DroneUniversity)

ISBN-10:
0692316035

1. The main category of the book —Aviation —Other category. 2. Another subject category —Robotics.

First Edition

V1.1 Updated links, fixed typos and added new info. 10/9/2014
V1.11 Updated links, fixed typos and added new info. 10/14/2014
V1.12 Print edition release. 10/17/2014

Preface

Note: This book is interactive! Select subjects include links to my website kiloOhm.com and other sites where you will find up to date information, additional content and more pictures. This book is meant to be used with the website for the full experience, please feel free to comment or ask questions on the site. This book is best displayed on a color e-reader to make best use of the schematics and diagrams, or view it with amazon.com/CloudReader on your computer! If you have the print edition, I will have shortened links that are easy to type into your computer browser.

Tarot 650 with 15" props hovering on autopilot. Several multicopter and flying wing recipes near the end of book. (http://amzn.to/1re1aBI)

Now on with the show!

So you want to build a First Person Video (FPV) model airplane or multicopter; AKA Unmanned Aerial Vehicle (UAV) and Unmanned Aircraft Systems (UAS), or just plainly named the good old *Drone*. While they have many different names, they all feature commonalities that have been scaled up or down to fit their application and budget. We will focus on designing an electric Radio Controlled (RC) model with a live video feed which is commonly referred to as an FPV, which in its most basic form is just a RC airplane or copter with a wireless CCTV camera on it. FPV is the mini experimental civilian hobbyist version of a drone and is not suitable for high risk uses.

To assist the reader, I have included several "recipes" near the end of the book. There you will find an inexpensive beginner FPV airplane recipe, quadcopters capable of carrying a DSLR or Cinema camera, a racing mini quad, and a recipe for a performance driven FPV/UAS wing that is capable of an estimated 15,000 foot ceiling, 1 hour flight time, 100mph straight and 120mph dive speeds, with a radio and video range of 40 miles. Sounds good huh! It's a challenge to get that kind of performance even on .gov budget, but it is possible to make a highly capable FPV UAV in your garage or on the kitchen table for a under a grand. Or you can scale down depending on your needs and budget and can build one for just a couple hundred dollars. You won't have to pour through hundreds of internet videos and dodgy forums to figure out what works, this book will highlight the parts required; expensive or cheap- your choice.

This book will break down all the required and optional components into six easy to understand sub-systems; Airframe System, Radio Control / Autopilot System, Camera System, Power System, Ground Control Station, and the Drive Train System. Other chapters will provide flying tips, check lists and a many surprises in-between.

Our focus is on engineering constants that will still be relevant far into the future. The aircraft of the future may look different than what we have now but will still operate according to what physics allows. However, as time presses on - components will get smaller, lighter, cheaper and more powerful which will increase flight times.

FPV 55" Ritewing ZII with 3DR APM autopilot and telemetry, Gopro 3, long range radios. Several recipes near the end of the book.

Table of Contents

Chapter 1: Overview

Before you get deep into the book; just know that all acronyms are defined at the back of the book. While you are there sure to check the back of this book for recipes for various custom drones.

Firstly I would like to clarify some terms. Since this hobby and industry is still in its infancy, people are still deciding on what to name these things. These definitions are based on the community understanding at the time of publishing.

Definitions

RC Airplane – Just a plain old Radio Controlled airplane.

FPV – A plain old RC plane, but with the addition of a live video feed.

Mini UAV – A FPV but with a semi-autonomous autopilot added.

UAV – Old generic term, autonomous unmanned air vehicle. Now called UAS.

UAS – Non Generic term. A UAV plus it encompasses all of the ground station equipment, training and documentation. Probably operated by more than two persons. Term is probably best used for purchasing reasons if the buyer wants a ready to run setup with support and warranty.

Drone – Unfortunately the term drone is blanketly applied to all things flying that don't have a human on board or RC. To some; it derives a frightening connotation that originated from the battlefields and hyperbole from the evening news. The military Predator drones are different, they have missiles and wage war, our hobby helicopters do not. But hey, the word "drone" it's catchy! I had to name my book this because it's the word of the decade. If it was named *FPV/UAS Univ.* or anything else, you probably would not have found it on the shelf.

Each of these vehicles are slightly different but they all root from the basic RC airplane or helicopter. You can take just about any RC airplane and turn it into a UAS if you want. There is a lot of flexibility, but ultimately you should figure out what you want to do and find which system will fulfill that best. This book will focus on the small FPV and Mini UAV video piloted variety that is operable by a single person, however you can scale it to however large or small you want and add as much robotics as you can afford to. The systems that are on a small RC aircraft and larger military UAS's are similar, and in some cases the same equipment is just scaled with larger engines, servos and power and redundancies.

Some duties that FPV, Drones and UAV's are performing today across the globe:

Cattle spotting
Aerial Photography
Wild land fire spotting
Forestry conservation
Search and Rescue
Archeology
Law Enforcement
Ski Patrol
3D terrain modeling
Construction
Navy and Coast Guard
Border Patrol
Agriculture
Infrastructure mapping
Meteorology
Fishing
Transportation
Movie and Television
Traffic Congestion
Oil Industry
Rail Industry
Military Surveillance
Zombie Hoard Control

The applications seem endless. If there is a manned aircraft currently performing a role, a UAV can likely fulfill that role or augment it, and if engineered well; excel at it too. In addition to current roles, I'm sure people like you have new ideas and will figure out new ways to use the technology as it becomes more available.

Safety

Just a few things before you get started, so you keep all the fingers you started with.

The technology described in this book demands respect: 500 mph tip-speed spinning propellers, electricity, radio emissions, kinetic energy. Don't try and catch a drone (yes, people have been cut-up trying this), and don't fly near people.

<u>The drones in this book are not toys. Do not buy one of these for your little kid.</u>

<u>Safety is always first! Do not take unnecessary risks!</u>

Knowledge

Your prior knowledge considerations: I will use some Electrical Engineering terminology with acronyms that are briefly explained the back of the book in the Acronym Definitions portion, or you can Google and ask the all-powerful internet overlords for help and research. Try the FPV related forums that I have linked at the end of the book.

In addition, it will help if you have previous RC experience. This book is meant to take the base knowledge of RC airplanes or multicopters and evolve it into a semi-autonomous drone. But, even if you don't have RC experience but you want a small drone, then this is the place to start.

Legalities

In the USA we have to follow the FAA guidance regarding Model Airplanes. Most guys at the FAA like to see neat stuff in the air and flying, they are just trying to fit the drone into existing airspace, so please work with them to help fit our toys and industrial UAV's into the new world.

Recently the FAA issued a $10,000 fine to FPV pilot 'Trappy' for flying near buildings and commercial use, he fought it and won on the basis that the FAA didn't have a clear law in place, but I'm sure they will now try to establish new laws. In other places in the world you might not have any laws at all regarding drones; New Zealand, Brazil and Mexico are a few that come to mind where it's unregulated.

In addition to the FAA, there are FCC laws regulating radio frequencies and power output. You can easily get a license (HAM) and be legal for maximum radio range.

Check your locale for other Federal, State, County and City laws and Home owners association laws and disapproving mother in laws.

Legislation

Become an advocate! If you see any civilian ownership anti-RC legislation in the news, please act and contact your Representatives and don't let them ban or further regulate FPV's or Model Airplanes.

There recently were rumblings of banning all RC aircraft due to Homeland Security concerns, while at the same time they are increasing the number of government owned drones in the skies over your head. Here is the basic reality: if a terrorist wants to fly a drone and kill people with it, there is no law that can stop him. A ban won't do anything to protect us, because a terrorist doesn't respect our laws. It's analogous to a criminal that runs around poking people with a *weaponized* sharpened stick. Banning or regulating all sharpened sticks in the country won't fix the problem, it will just inconvenience law abiding citizens, stifle stick sharpening technology and turn good citizens that use sticks for roasting marshmallows into, in fact- criminals.

Things like legislation can directly or indirectly affect technology negatively. Thankfully all of the FPV systems crossover into consumer electronics, so replacement parts are not that difficult to source. For instance; the entire video broadcast system on the plane could be interchanged with a simple TV baby monitor. Even a regular old cell phone can replace a purpose built camera and video radio. If this is all you had, *or are allowed to use*, things can easily be repurposed for FPV use. Bans on physical items are futile and only appease people who likely haven't the capacity to think things through.

Join the Academy of Model Aeronautics and write your representatives when needed.

Mantras we will build on

Lightness and efficiency are paramount – These are the keys to high performance in all aspects of aeronautics. All high performance aircraft is dependent on this simple concept and should be the defining principle in your design.

K.I.S.S. (Keep It Simple Stupid) – Simple engineering is generally better. In the case of aeronautics; simplicity could also equate to lighter overall weight which equates to a faster plane or more endurance. And with minimal systems in place there is less to go wrong which means you will have more fun time in the air and spend less time making repairs, theoretically…and dependent on flying skill.

"Modularity" – Is a concept that we will strive for to future proof our systems and to reduce one-off engineering and for ease for finding replacement parts. Open source hardware is usually preferred because it is vetted and popular and has longevity. For instance the Ardupilot system has hundreds if not thousands of people working on it, and if you have a tech question to ask I would rather ask thousands online than a single company that makes a widget with one or two people working in their support system.

"Two is one, and one is none" (*optional!*) – An old wise saying that states that your system is only as strong as its weakest link. The weakest link might be unknown until it goes out, so it's good to have backups. This might sound contrary to KISS because it adds complexity, but think of this as KISS with a backup, and in some scenarios this is the only way to go and in other cases absolutely not the way to go. If you want a fast FPV park flyer then disregard this saying because there is no need for backups and will only slow you down.

A difference between hobbyist and professional unmanned aircraft is the addition of backup systems. If you want to build a real deal UAV it should have redundancies; everything should have a backup. One simple way to avoid single point failures is to place double of all the normal RC components and have them in parallel independent of each other; essentially two planes worth of electronics and twin engines, switched by the pilot on the ground and/or automatically transitioned onboard. This would add twice the weight and should only be done on a large wing that can support that type of payload, additionally it's twice the cost and should only be employed when the UAV is required to perform a critical mission like carrying a RED HD camera, which we would use a Octocopter that can survive a failure of up to two motors. Consider this *insurance* when you are carrying an $8000 camera.

Finding Parts and Estimated Costs

Included throughout this book are web links that will help you source parts to build a quality FPV airplane or 'copter. Also provided are alternatives for components in case a supplier is out of stock, or sourcing and making components from everyday items. You will be free to choose to make a professional or a hardware store drone or something in-between.

Included is a complete "recipe" build list near the end of the book for a proven high performance FPV system.

The overall costs for a complete FPV system can wildly range depending on what features you choose and what you already have on hand. Some of the more expensive items required you may actually already own (a TV and laptop). A home improvement store foam drone will be inexpensive with minimal components, but if you are inclined to build a long range carbon fiber airframe with all the cool stuff, the costs go up.

Estimated Airplane UAS Costs

Airframe, 50-90" Wing: Home built or mail order. $20-175

Onboard Radio Control: Servos, autopilot, transceiver(s), GPS, telemetry/OSD, antenna(s). $200-500

Onboard Video System: Camera(s), video transmitter(s), antenna(s). $40-500

Power System: Batteries, connectors, chargers. $100-200

Drive Train: Prop, prop adapter, motor, speed control. $60-150

Ground Base Station: All ground radios, antennas and video monitors, laptop ($0 most already have this). $130-500

The 55" flying wing UAS build list at the end of the book was built for only $876! Including all FPV radios, batteries, chargers, monitor, everything even the shrink wrap and glue!

Use quality components. If you go cheap with no-brand radios and cameras that don't have a track record, you may suffer with poor range, video quality and reliability which puts your plane at risk. I have listed in the systems chapter's some quality components that are currently used and in good standing in the community.

Estimated Multirotor costs

Check out end of the book for the 5" prop Mini-H quad build and 15" prop giant quad build! $500-1500

Chapter 2: Airframe System

For our purposes the airframe is simply the wing or fuselage/skeleton/frame, it's the structural chassis that holds all your components and provides lift.

To Buy or Build?

Buying an airframe for your project will be easier and quicker to get flying and has considerable advantages over a home built. With a pre-built frame, the aerodynamics have been pre-sorted for you and all you have to do it place your components.

The popular RC planes selling now are made of very strong and durable EPP or EPO foam, they are fast and efficient. It is also possible to build a flying wing from that pink or blue foam insulation board (EPS) that they sell at home improvement stores. The disadvantage to making your own wing is that you will have to build prototypes and devote time to optimizing shape and size. Due to the one-off homemade experimental design you may also crash at a higher rate, risking your expensive electronics. My point of view is to initially use proven systems and not to over complicate things, simple is better and more reliable. I also don't see much logic in placing thousands of dollars of electronics in a ten dollar home built wing. Wait until you have experience before you attempt experimental designs.

ARF & RTF available Hobbyking Quanum Nova also called the Cheersun CX-20: inexpensive pre-built mini-APM 2.6 based quad. Good starter quad but still has advance features like waypoints and RTH. Don't buy the "Zero" version, it does not have a APM.

Common Platforms

Flying Wing

I prefer flying wings over conventional airplane styles because they are very efficient and simple. All surfaces are providing lift with no drag from fuselages and tails. But they are not without their issues, flying wings may exhibit some yaw wobble (a left/right wiggle on the center axis). To counteract yaw wobble; "winglets" are added to the tips of the wings. Yaw wobble can be completely removed with some tweaking and balancing. They are also tricky to launch.

Another consideration is that flying wings are Pitch sensitive and the Center of Gravity (CG) should be tuned or it will exhibit excessive stall, or will dive or stall aggressively, more-so than a traditional plane that has a tail with elevators that can counter trim a CG imbalance. CG is also closely related to battery endurance, due to fighting the trim with too much acceleration.

Flying wing build characteristics: Efficient and lightweight. Takes some skill to assemble and install components. Shrink covering and spars are required for strength.

There are few manufactures of wings, the **RiteWing Zephyr II** 55" has EPO foam which is durable and flexible $130, they also make a wire cut Z60 made of EPP foam. Available in everything from 34"- 90". Also there is a new wing called the Venturi FPV by Flyingwings.com, and a new Zephyr III from Ritewing.

55" Ritewing ZII, that's been through about twenty major nose dive crashes. Got a new nose on her.

Blended Wing

Skywalker X8 - Easy to get going. 2910g, 2120mm. Larger payload area than a flying wing. Technically a "blended" wing due to the bulbous payload area. Made of uncovered EPO foam, but covering with a laminate would add some additional strength. Hobbyking.com or BevRC.com are reputable importers of it. $175

Skywalker X7 – Newer and sleeker than the X8, but nobody knows much about it yet. Poor marketing I guess.

Rangevideo RVJet 1950mm. A new EPO blended wing that is easily deconstructed for transportation. Comes uncovered, so at least add packing tape to the leading edges. $250

Homemade Wing

Moderate to hard difficulty to build. Inexpensive and locally sourced at home improvement stores. Usually made of pink or blue insulation, EPS or XPS foam which is brittle but works. You glue the slabs of pink or blue XPS foam together then trace out the template and cut it with a 'hot-wire'. If you are lucky enough to find a piece thick enough so you won't have to glue them together, then the foam is cut to shape with a hot wire. A hot wire is just a piano wire hooked to a lantern battery. Plans for planes and hot wires are available online. May not work correctly the first time out and require lots of fidgeting.

Another model plan is called the FPV49. It's made of thin slabs (½") of Depron foam and looks easy to build. Check it out online.

If you want to add spars for rigidity, just use a fiberglass arrow shaft! They are strong and light, I use the Easton Scout 2 that I got at Cabella's. They come in lengths up to 25" and can be found at any sporting goods store. 6 pack was $19. They are 6.66mm od with 4.86mm id and are orange in color. After cutting off the tip and stripping the fletching, it is 64.5cm long and weighs 11g, nice and cheap and not RF reactive.

Non-Flying Wings, Conventional Aircraft

Pushers – These are RC planes that have the propeller in the rear but not a flying wing. They have a fuselage. This is also a good choice for FPV as long as the propeller is located in the rear so that cameras are unobstructed. The original

Skywalker and *Bixler* are/were popular models but were not designed for FPV like the new twin-tail *Skyhunter* that has lots of payload space for cameras and the *Techpod* deserves mention as well as the new *Bix 3*.

You can add FPV to any RC airplane! – Fly whatever you want! Some people prefer to have the spinning propeller and faux cockpit in the video, gives a sense of flying a real airplane. A twin or quad engine WWII bomber or a P51 Mustang would be a lot of fun. Just take into account that they might not be the most aerodynamically efficient, so long range flights shouldn't be attempted with these.

Rotorcraft

Helicopter – The good old flying lawnmower. Seriously though, people have been accidentally decapitated by large RC helicopters. More dangerous than multirotors, due to the large blades that have much more torque. Also since they have several gears and bearings and moving parts, they require more maintenance.

Coaxial copter – A helicopter with two couter-rotating rotors, one on top of the other.

Bicopter – Not really a thing yet. It's technically difficult to stabilize without pendulum'ing. It's going to be while until you can have an *Avatar* style Pandora gunship.

Tricopter – Three props in a Y configuration, fast and agile and low cost. The rear motor has a servo to steer like a rudder. Checkout the Fortis Titan-II.

Quadcopter – Four props. Available in X shape, H shape, K shape aka Deadcat. H and K are best for video because it keeps the props out of the video. Checkout the Tarot 650, Blackout Mini-H, 3DR Iris+ (http://amzn.to/1re29C7), Hoverthings Flip FPV, Dji Phantom (http://amzn.to/1oKGcuB).

Y6 – A tricopter frame but with 6 props, gains some moderate redundancy but has efficiency losses due to turbulence of the stacked props. Also louder than a hex. Checkout the 3DR Y6.

Hexacopter – Six props, moderate redundancy. Might be able to limp with one motor failure. Checkout the Tarot 680Pro, Blackout mini-Hex, Tarot T810, Tarot T960.

Octoquad – A quadcopter frame but with 8 props, gains some redundancy but has efficiency losses due to turbulence from stacked motors, sort of like 4 coaxial copters tied together. Smaller footprint than an octocopter, but louder and less efficient than an octo. Checkout the 3DR X8.

Octocopter – Eight props. High redundancy, but twice as expensive. Can fly if two motors fail, as long as the two motors that failed are not near each other. Can carry large payloads, DSLR's and RED cameras. Check out the Tarot 1000mm, Gryphon Dynamics, Dji S900.

Decacopter – 10 props, **Dodecacopter**/duodecacopter– 12 props (Gryphon Dynamics), **Hexadecacopter** – 16 props. There are very few of these *rarecopters* in existence but expect to see more in the future as payloads get heavier.

Tip: Multicopter frame size is measured from the farthest motor to the closest motor in millimeters. Show above is a homemade 250mm mini-H quad.

Materials

Modern RC airplanes are mostly made of foam and corrugated plastic, but some are still Balsa or Poplar wood. Multirotors are carbon fiber, glass fiber or plastic and even sometimes wood.

Foam Types

EPP – Expanded Polypropylene foam. Commonly used in electronics packaging, this is the type of foam that your hard-drive or a new laptop computer is shipped with. Durable, flexible. Handles impact very well. The best choice, but very few planes are made of this new foam.

EPO – Expanded Polyolefin foam. Similar to EPP but not as flexible. Stiff, durable. Handles impact well. Second best, most modern foam planes are made of this.

EPS – Expanded Polystyrene foam. Think disposable coffee cups, meat trays and surfboard foam. Brittle, very lightweight. Dents or breaks on impact.

XPS – *Extruded* version of EPS. Blue or pink foam board insulation that is available at home improvement stores. Brittle, lightweight. Dents or breaks on impact. Also called Depron.

Foam wings are recommend to be covered with shrink covering, and put packing tape on leading edges. Helps in durability for crashes.

Foam is very lightweight and easily repaired. Wings will require support spars to add rigidity. Spars should be preferably fiberglass, not carbon fiber or metal. Carbon spars might mess with the antennas because it is conductive, they could interfere or block the antennas. The jury is still out on how much or if any effect there is from a thin carbon spar.

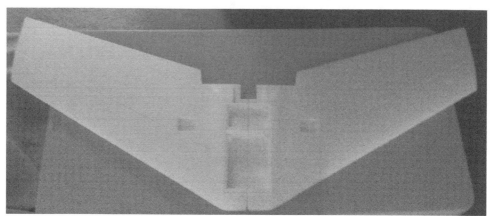
Ritewing ZII in EPO foam.

Corrugated Plastic (trade name *Coroplast*) - An extruded plastic product, similar to cardboard but made of plastic. Used for battery covers, winglets and rudders and entire tail sections. Found in election signs and "for sale" yard signs. Here is an Amazon link (http://amzn.to/11kx22N) that is a source of multicolored corrugated plastic if you don't have any politicians or yard sales in your area.

Composites – Instead of foam, you can build one out of a light fabric and resin but don't bother working with raw composites and resin to cover airframes until you have a lot of experience. PITA factor goes up greatly when dealing with laminated composites. But if you like pain: If you want a carbon fiber wing you can use a "lost foam" method, you cover it with a couple layers of 3k fabric then vacuum bag. Once it dries, pour gasoline into the foam to melt it. You will be left with a carbon shell. Sounds janky but it works. The ideal way would be to use carbon prepreg and vacuum bag then autoclave it, the final shell ends up being lighter due to less resin used.

> **Carbon fiber fabric (CF)** - Extremely durable and lightweight. <u>Carbon *weave* will interfere with antennas and block signals.</u> Antenna placement becomes troublesome. Difficult to build and repair.

> **S-glass, light weave fiberglass fabric** - Won't interfere with radios. Semi-Durable and lightweight.

> **Kevlar fabric** - Won't interfere with radios. Most-Durable but relatively heavy. Difficult to build and repair.

> **G10 / FR4** – Fiberglass and resin laminate. Most circuit boards are made of G10.

22

Micarta – Sort of like G10 but made of cotton and epoxy. I made my own mini-H quad frame pieces with 2 layers of old jeans, one carbon and one Kevlar then 30 minute epoxy with a brick on top as a press. Used wax paper as a release. Works great and is very strong. It's not my first choice in a material, however I made it out of scrap junk in about 35 minutes which beats UPS any day.

Wood – Balsa...I hate this stuff, but I guess it needs a mention. It's so weak...so sooo weak, but very lightweight. Use poplar wood or foam or Coroplast if possible.

Consumables

Gorilla Glue – This is polyurethane glue is for gluing all foam types listed above, and mounting spars and etc. Apply glue then activate by lightly spraying with water. http://amzn.to/1167XT4

Scotch Extreme Tape – Reinforced. Good to use as aileron hinges and to put on leading edges for landing durability. Or use Storage tape. http://amzn.to/19A8UsI

Kapton Tape – Polyamide 500 degrees heat resistant and non-conductive. I wrap electronics in this and in some places use it instead of cable ties because it's lighter. http://amzn.to/1s37TDT

Loctite Blue Threadlocker - Every screw should have some of this on it. (as long as there is no plastic, Loctite weakens plastic) http://amzn.to/170FktO

CA – Hobby shop Cyanoacrylate, it's similar to Superglue but faster drying. Will erode foam. http://amzn.to/ZPq2uo

CA Activator / Accelerator– makes setting quicker, but may weaken the bond. I don't like to use this. http://amzn.to/1tsZHJw

Cable "zip" ties – Use the UV resistant outdoor nylon ones, they are the best. http://amzn.to/1oKJiyJ

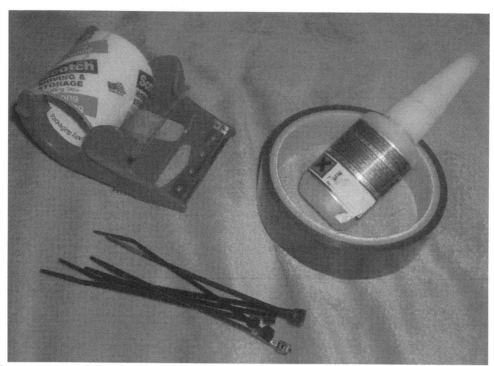

Various consumables that I use the most of. Storage tape, Kapton tape, CA and cable ties.

Chapter 3: Radio Control & Autopilot System

This chapter will discuss the associated parts that control the airplane; the human with a controller or the autopilot. You can use just a RC system for a basic FPV, or you can add an autopilot.

RC Frequency Selection

There are several frequencies to choose for your RC transmitter and receiver by swapping "Piggybacks" onto your RC transmitter. For long range choose 443MHz. Be sure not to choose the same frequency that your video transmits on. I recommend 433MHz 500-1000mW for RCtx and 1.3GHz 500-800mW for Vtx long range. For short range; feel free to use 2.4GHz or 5.8GHz or really whatever you can get your hands on

Power levels should be kept under 36dBm EIRP total. If you are using omni antennas, the gains will be low on the antennas so you could technically use 4W of power, which is a whole lot. If you use a ground station with a directional helical RCtx antenna that has 9dBi gain you would then only be legally allowed to use a 500mW transmitter for a total of 36dBm EIRP (or 4W of directed power in this FCC band, is another way to think of this).....these are of course FCC laws and will vary where you are located in the World.

Here is a list of RC associated frequencies available:

27 & 35 & 72 MHz RC, These are "old style" transmitters from the 1980's with the crystals in them. Technically due to the lower frequency they should go very far however only use them for short range. Don't use these for anything other than park flying, because the transmitters are low powered and don't frequency hop.

433MHz (UHF) LRS (*long range system*, RC community colloquialism, that only the most awesome people know about. Welcome to the club!). These LRS's are only available as an add-on piggyback that mounts to the back of to your modern RC controller like Taranis, JR, Futaba, Spektrum or Turnigy. Signal travels around trees and buildings pretty well. On low end cheap-o LRS's an additional low pass filter may be necessary due to the 3rd harmonic: 433 x 3 = 1299MHz, use an SDR to check. Oh, and you should be a HAM to use this frequency at these powers.

Flytron OpenLRS - LRS 433MHz, 1W. 30-40 miles est. $145 for RCtx and RCrx, firmware upgradeable and customizable. Requires flashing with 3V FTDI. Telemetry downlink possible. I've tested with 9XR and 9XRpro.

Hawkeye / DTF UHF – A preprogrammed Flytron clone. LRS 433MHz 1W 30-40 miles est. $90 for RCtx and RCrx. Built in low pass filter, good for FPV. No antenna incl. Telemetry downlink possible. I've tested with 9XR and 9XRpro.

OrangeLRS Hobbyking – "LRS" 433MHz. 100mW or 1W power. 1-3 miles, 30-40miles. Requires OpenLRS firmware programming. Reports of reliability and build issues, I would use a Flytron or Hawkeye over this.

RangeLink – LRS 433MHz, 200/500mW. 10-40 miles est. $185 for RCtx and RCrx. Purportedly a Dragon link clone. I've tested with 9XR and 9XRpro.

Dragon Link – LRS 433MHz, 500mW. 30-40 miles est. $269 for RCtx and RCrx.

ImmersionRC EzUHF – LRS 433MHz, 600mW. 30-40 miles est. $200-310 for RCtx and RCrx.

Scherrer TSLRS – LRS 433MHz, 200mW to 8W boosters. 10-60+ miles est. $450+ for RCtx and RCrx.

2.4GHz (2370-2450MHz) - Fly behind some objects and a few trees. Lots of WiFi environmental interference, but these frequency-hop so it should be okay. Typically 100-250mW. 2-4 miles est. To get more range, use a LRS add-on listed above and/or directional antenna, 15 miles est. Most 2.4 systems are just one way data uplink and there are tons to choose from, but the FrSky has a telemetry downlink available. And you don't have to be a HAM.

Good stuff: Lemon DSM2 receiver, FrSky D series for telemetry

5.8GHz RC (rare) is only acceptable for a "park flier", trees and buildings completely block signal. Dji is the only one that uses a 25-125mW 5.8 RC transmitter for their Phantom 2 Vision. (http://amzn.to/1vIBzFu) This is so they can send video over 2.4GHz WiFi to your iphone. The rest of the industry uses 5.8GHz exclusively for video. 300m range after enabling FCC mode.

Flytron OpenLRS *transciever* 433MHz internal backpack module in a 9XR. Little known secret; if you buy two, use one of these *on the airplane*. You can then enable 1W telemetry with the proper firmware. Note that the slot cut on the left to allow access to programming pins. 3V FTDI – USB required.

Rangelink UHF Receiver

Rangelink 433MHz LRS external transmitter.

Antenna Styles for onboard RC

Since our little FPV drone won't be able to feature a tracking-steerable **uni**-directional dish antenna due to weight and expense, we will have to use an **omni**-directional for all antennas mounted on the aircraft because the aircraft is moving.

The base station can feature either uni or omni directional antennas, uni is recommended for longer ranges steering the base station antenna is necessary depending on how tight of a focus your antenna serves. Example: If you are in a park flying around in circles around yourself, use omni antennas on your base station. Or if your goal is long range then use a uni-helical on a tripod that you can manually steer.

A uni-directional antenna is focused like flashlight that can reach out long distances but only one direction, and omni-directional is like an end table lamp in that the light goes in (almost) all directions.

Polarization – we use linear polarization for RC, and if possible- circular polarization for video, which will be explained in the video antenna chapter.

These are the common styles that are used for the **onboard RC**.

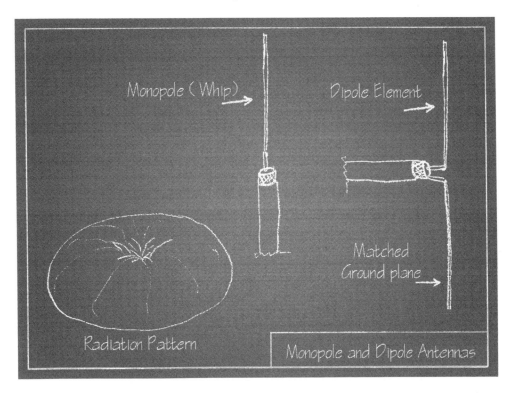

Whip Mono-Pole and Dipole – Classic omni-directional antenna. Pictured center is a "Whip", which is half as efficient as a dipole. Both have single projected donut, which can negatively affect your signal when banking with dark zones on top and bottom of the element. Obsolete for the most part for video due to linear polarization and a single donut lobe, the Skew Planar listed below is better suited for FPV. I use homebuilt dipoles for telemetry and RCrx because I am not worried about multipathing with RCrx.

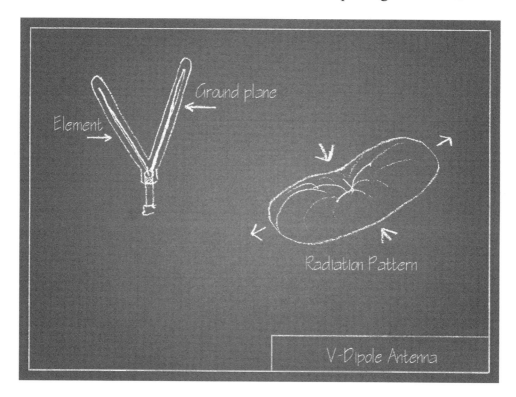

V-Dipole –Shaped like a V and was the standard for FPV. This is a linear polarized pseudo-omni that features *a modified elongated single* donut shaped lobe that give good coverage in two opposing directions and fair coverage in all directions. The construction is simply a bent dipole that one arm is connected to the ground and is used to deform the donut lobe. But since we now have circular technology now, it's mostly obsolete. But if you are not concerned with multipath interference then the V remains a good choice and will give more gain in the direction of the two pointed lobes than an omni Skew Planar.

Additional Antenna Considerations

Notch filter – Not an antenna. This is a device that attaches between the antenna and receiver, on both the plane and/or the base station. They are tuned for a specific frequency and their purpose is to block that one frequency or you can use multiple filters to block multiple individual interfering signals. Use when you have a known specific interference.

> Example: 433/1280 filter placed on a 433MHz RC UHF LRS transmitter will reduce interference on a 1.3GHz video receiver.

Ibcrazy 433/1280 Notch filter. The "arms" are actually finely tuned 1280MHz antennas that are parasitically grounded, so they capture and trap 1280 and don't let it pass on.

Low Pass filter – Used to "turn down the volume" of frequencies **above** the filters rated freq. Example: 1.3GHz Low Pass filter placed on a 1.3GHz video transmitter will reduce interference on a 2.4GHz RC receiver. Quality transmitters will have an integrated low pass. One problem with these is higher insertion loss than a notch filter.

High Pass filter – Used to "turn down the volume" of frequencies **below** the filters rated freq. Example: 2.4GHz High Pass filter placed on a junky 2.4GHz RC transmitter will reduce interference to receivers below 2.4GHz, such as the GPS. It should only be needed if using a low quality transmitter that spews energy downward outside of its rated frequency bandwidth.

31

While using any filter is a quick fix, it isn't always the best way. Since a filter also reduces dB of the whole system slightly (insertion loss), a better way might be is to place the components further apart to logarithmically reduce interference. I had issues with my 433MHz stomping on my 1.3GHz video on my base station, to fix it all I did was move the Vrx antenna eight inches away.

Impedance – Most if not all FPV and RC radio gear has a standard of 50 Ohm impedance. What does that mean for you? It means that you should match your impedance to your transmitter; so your transmitter output is 50 Ohms then select an antenna and cable that is 50 Ohms.

Autopilot & OSD

The Autopilot system includes the accelerometers, IMU (compass), GPS, airspeed sensor, temperature and barometer sensors as well as other optional sensors. The radio receiver's signals are routed to the autopilot where sensor considerations are made and then data is outputted to the servos and speed controllers. A true autopilot will be able to fly waypoints, auto land and takeoff, loiter and return to home. An autopilot can't compensate for broken physical parts such as ailerons, stuck servos or other malfunctions. Which is why you should be in line of sight or have a video link up.

The on screen display (OSD) is technically part of both the video system and the autopilot system. The OSD is data that overlays your video with speed, altitude and battery usage amongst other data. An OSD is a piece of hardware that fits between the camera composite output and the video transmitter input and a data line going to the autopilot via mavlink protocol.

Autopilots have onboard accelerometers for stabilization that keep things level and remove unwanted movement and compensate for turbulence. Your plane will be much more stable while flying in winds. This is particularly important when using a "return to home" system because sometimes it's easier for the plane to fly its self. I've flown just fine in 30 mph winds, well actually that's a lie. The Autopilot was doing the flying.

Here are some top gun **Autopilots and "dumb" flight controllers,** and a few not so great ones. There are several different makes on the market, it can be daunting.

3D Robotics ArduPilot APM 2.5 / 2.6 – A popular full featured autopilot that can used on planes, quads or ground vehicles. Open Source 8 bit Arduino based. Telemetry can be added; 3D Robotics 900MHz 100mW but you won't get far on that without a booster. Ground station software is available and free. Large forum community. Requires current sensor minimOSD and GPS UBlox-6 and optional airspeed kit. $75-300 total depending on origin. Beware of poor clones out there. V2.6 includes an external compass (magnetometer) so it can be placed away from motors. This is what I and NASA like to use because its old/proven. (http://amzn.to/1vIyRA3) (http://amzn.to/1q61GBA)

3D Robotics PX4 (Pixhawk) – New successor to the APM, features a faster 32bit processor and more memory. With this you can also run a gimbal control onboard. $200-$280 (http://amzn.to/1xiOQFQ)

3D Robotics / Intel Edison – Next gen we can look forward to. So tiny and powerful, this is going to be great.

Paparazzi – A true full featured totally open platform similar to PIXHAWK's capabilities in many ways. The Lisa/S model has built in Ublox gps and its super tiny! *Hak5* has a few really excellent podcast episodes which walk you through the whole Lisa/MX setup. $230-280 + $60 JTAG programmer.

OpenPilot – Another open platform with a strong community. Has everything that APM and Papa does. CC3D is the older small board but with 32 bits it's still great for small quads and still works great, won't have GPS just the basics $20. The Revolution 32 bit set has a faster processor and includes GPS and telemetry for $200, quite a deal indeed! (http://amzn.to/1uu30Vf)

uThere Ruby – Like the others above this has auto take off, assisted flight/fly by wire, loiter, autonomous landing, panic mode/RTH. Comes with its own GPS, current sensor, barometer, airspeed, and magnetometer. The OSD is an add-on module (RubyOSD) – $140. Under development is a new ground control software with way-point generation for PC and two way 900MHz telemetry at 250mW and a *750mW!* This could make for a complete robotic long range UAS. $345 + optional OSD & telemetry systems. Airplanes only and closed source hardware and code. Updated info 9/29/14: "uThere is shifting away from sales of components to individual hobbyists and focusing on working with manufacturers of "ready to fly" systems such as RitewingRobotics.com or AgEagle.com" - Jim from uThere.

MegapirateNG – Ardupilot/copter/rover port. Supports many different cheap 8-bit hardware boards such as KK or Crius AIO. Now that APM hardware is inexpensive I don't see why you would use this but maybe if you already had stuff laying around it would be a good project.

VR Brain - another Arducopter port, 32 bit.

Other autopilots, but for copters only: Dji Wookon $1000, DJI Naza (after waypoint firmware update) $300, XAircraft SuperX $400-500 (http://amzn.to/1vItgcO)

'Flight Control' boards for copters: These are 'dumb' in that they don't have waypoint control or other advanced features like autopilots, they generally only have accelerometers and a barometer and maybe a GPS for a very basic RTH and loiter mode. They are only for general stabilization of quadcopters by control of the ESC's. This is mentioned because they can be directly replaced by a full autopilot and are often called autopilots, although they are not. They are really inexpensive and are a great way to control a basic RC quadrotor, $20. Examples: Lemon Flip FC/RCrx, KK2.x or MultiWii or Naze32 (I love my Naze32!), Dji Naza, AeroQuad. (http://amzn.to/1BJNsNf)

Add-on OSD, no autopilot included. – An OSD is device that interfaces with your video transmitter and overlays HUD information, like speed and altitude or a hundred other items that you want to see on your LCD monitor at base camp. Some like EzOSD have their own sensors like GPS and accelerometers to provide the data, others like MinimOSD don't have sensors and rely on the autopilot for that information via the MAVLink protocol. OSD's won't have automation like waypoints.

> **MinimOSD** – An inexpensive OSD made for APM or Pix or others, $40 for the original $15 for the clones. Lots of guys burn them up because the units can't handle the 12 volts. Wiring is not straight forward, it has a 12V input, but do not apply 12V to it, just apply 5V. Look in my schematics section of the APM plane for correct wiring. And follow my firmware upgrade to Extra. (http://kiloohm.com/?p=61)

> **EzOSD** – An add-on OSD with GPS, Current sensors and other sensors. Has one way audio channel telemetry downlink with iPhone and Android apps. Next best thing to an autopilot, but spendy. $180

> **Others:** Eagletree OSD, Dragon OSD, Rangevideo RVOSD.

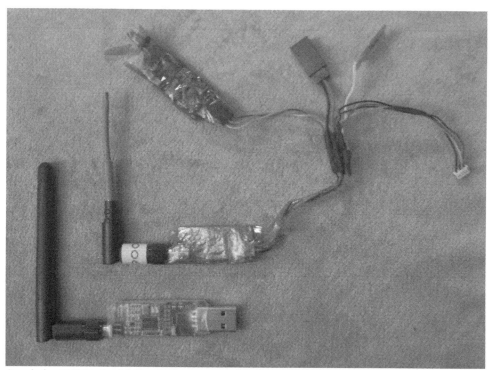

(top) <u>minimOSD</u>, (middle and bottom) <u>3DR 900MHz telemetry</u>. Telem antenna is de-cased for lightness, and the board is now shielded from noise with <u>kapton</u> and aluminum. Wiring diagram at end of book. Also be mindful of the antennas are actually what you ordered. I bought some 900MHz telemetry systems from China and they arrived with 2.4GHz antennas. (http://amzn.to/1q61GBA) (http://amzn.to/1vIyRA3) (http://amzn.to/1s37TDT)

GPS Modules

This is a relatively short section because these modules are mostly plug and play with your autopilot. Just make sure you pwr, tx, rx and gnd wires are correct and the baud rate is set in the gps's firmware. With APM and Pixhawk systems using uBlox, make sure that your default uBlox is set for the higher baud rate. Most ship with 8800 baud, they need to be configured with the 3dr firmware with a baud of 38400! http://copter.ardupilot.com/wiki/common-ublox-gps/

Be sure to mount the module somewhere on top of the aircraft with a clear view of the sky and away from the transmitter and transmitting antenna and motor and before flight, make sure your GPS has a good lock, takes about 30 seconds from power on. Reception of 6 satellites in the minimum, 9 is normal.

There are many factors that make a GPS good or not, I have listed just two main factors: boot time in seconds and accuracy in meters which is what we care about.

Popular GPS Modules

uBlox LEA-6H – full options available in firmware, warm starts fast, tracks very well. (26sec warm/cold start, 2.5m accuracy). $50

uBlox NEO-6M – a cheaper version uBlox, but works good. 27s and accuracy of 2.5m. $15-20 (http://amzn.to/1rSFqz5)

uBlox UBX-G6010-ST – new low-cost chip with Neo6 like performance. Rctimer. $25

uBlox 7 & M8 – New chips. I haven't used yet but want to. M8 has 72 channels!! (29s, 2m) VRX GPS NEO-M8 or MAX-M8 are on the market. $75 (http://amzn.to/1xXZAwI)

Mediatek MT3329 – 3dr's GPS (34s, 3m) $38

Globalsat EM-406A – Older tech, also not cheap yet for some unknown reason. (42sec cold start boot, 10m accuracy... yuk) $40

Globalsat EM-506 – (35s, 2.5m) $40

Ublox Neo-6M chip on a Crius board and antenna.

Servos

Sizes – 34" planes will use micro-servos, 50-60" planes will use standard servos, 70" planes might use a larger servo or multiple standard servos.

Digital or Analog? – Digital servos have a microprocessor chip that is analyzing the input signal for noise. Get Digital if you <u>need</u> an accurate flying plane, such as a 3D plane or heli. Digital will be more precise, but it uses more battery. If you are going for endurance or any other reason, then choose analog.

Grams – This is the class of the servo and also the weight of a servo and only a GENERAL indicator of its performance. Example: 30"wing uses two 9 gram servos, 55" wing uses two 45gram servo, 70" wing uses four 45 gram servos in concert.

Ball bearings, gears, bushings – Metal gears and bearings are superior to plastic gears and bushings, but cost 4x more.

TIPS:

-If at all possible choose a servo with bearings and not bushings, less play and last longer.

-Your control surface throws should be about 8 or 10mm up and down for the flying wing builds.

-Some good servos: Hitec MG-645, <u>Hitec HS-645MG</u>, Goteck gs-3630bb. (http://amzn.to/1uu2Kp5)

-Wrap your servo in electrical tape before hot-gluing into a foam wing, it makes it easier to remove later.

-Don't use regular craft store crap hot-glue, it melts at 140f and on a hot day in your car will make a mess. Use "hair extension" <u>polyamide hot glue</u>. It melts at ~375f. It's cheap on eBay or Amazon and your new hair will look great too. (http://amzn.to/1rdWKuC)

-Also, Glue "guns" are unnecessary and messy!!! Just heat the tip of the glue stick with a lighter to soften and apply like lipstick, nasty black burning lipstick. So there, you now have new hair and lipstick. Great job!

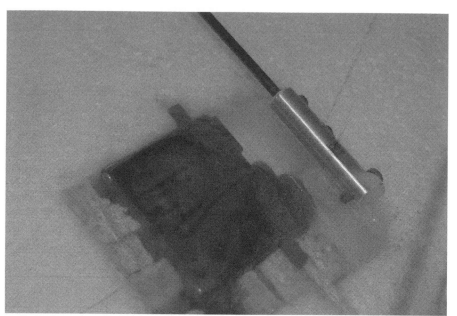

21 Gram digital servo, fit into a 45 gram servo slot with balsa spacers and glue. Used this servo because it was as strong and quick as the 45 gram it replaced but lighter and more accurate.

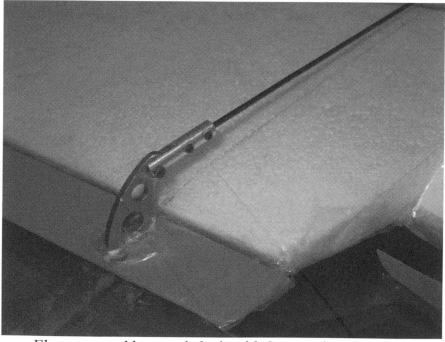

Elevon control horn and clevis with 2mm carbon fiber rod.

Chapter 4: Camera System

There are two schools of thought. Most FVP'ers will use a tiny flight camera for flying and a secondary HD camera that's recording for offline viewing later. The other way is with a single camera and use it for both recording and flying. The later isn't as safe because mini HD cameras have a tendency to auto-power off or seize up. Use a reliable dedicated 1/3" FPV flight camera and record HD on a second camera.

Flight cams, 1/3 inch CCD

Flight cams are usually 12 Volts but some are 5V, so a converter/regulator may be needed. I use the Adjustable Sepic up/down regulator. You also might need to add a LC power filter wired inline nearest the camera if you have any interference from your main battery/ESC or use a toroid. On some larger planes I would use a separate battery for video and RC all together to eliminate interference and provide redundancy. Wrapping your cameras electronics in aluminum, copper foil or carbon fiber might help too; a lil-faraday cage. But if done improperly will lead to poor results and could make it worse. (http://amzn.to/1sE5Aat) (http://bit.ly/1sIhaTa)

Good FPV flight cams
Sony PZ0420 – 600tvl, .01 lux $30 http://amzn.to/11kEcEc

Sony CMQ1993x - 600tvl, .01 lux $30 http://amzn.to/1sdokwb

Sony 540TVL - .001 lux capable, not *real* night vision but close and will work if you have some passive lighting. $50

Fatshark Predator v2 camera kit– camera comes with a 5.8GHz transmitter, easy and tiny and comes with video glasses. Good option noobs and for small short range craft and ease of use. $279

*GoPro and Mobius may also be used with video out cables, however they have some processing lag and could also power off mid-flight leading to a crash.

Lenses, which one to choose for my 1/3" CCD camera?

1.78mm = 165° viewing angle, Super Fisheye, widest view. Good for quadrotor, skateboard videos. Has some vignette (black corners of image).

2.1mm = 150° viewing angle, Semi-Fisheye. Good for quadrotor, not airplanes.

2.8mm = 86° viewing angle. Good for airplane FPV.

3.6mm = 65° viewing angle. Good for airplane FPV.

6.0mm = 50° viewing angle. Only for mounting on a gimbal, narrow view.

…and so on, the lenses get more telescopic and would require active stabilization with a brushless gimbal.

My camera needs 5V what do I do?

-Easy! plug a LC filter into the BEC/speed controller's 5v output and plug your camera into that.

My camera needs 12V what do I do?

-If using a 2S battery or smaller (<7.4V) Use a switching **UP** DC-DC Sepic type converter to change voltage to 12V. And a L-C filter to smooth that voltage.

-If using a 3S battery (11V) you can run the camera straight off the pack or balance lead. Most 12V cameras will work down to 10V.

-If using a 4S battery or greater (>14.8V) Use a switching **DOWN** DC-DC Sepic type converter to change voltage to 12V. And a L-C filter to smooth that voltage.

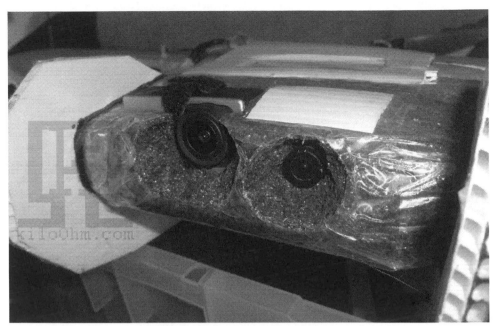

GoPro 3 Left, PZ0420 Right.

PZ0420 covered in dirt and grime. This quad has led a double life as a lawn dart, but because the camera lens is recessed it has survived. Yes the frame is ugly, it's experimental and is allowed to be. ;)

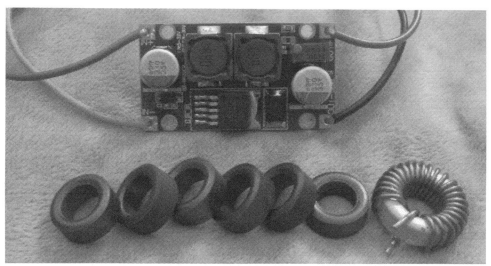

(top) Sepic type DC-DC voltage down converter, (left) toroid cores, (right) single wire toroid.

HD Camera

A small sports type HD cam will be more than enough for most people, and can be as cheap as $75. But if you need a better picture then scroll down for larger format cameras.

There are several types of mini HD cameras on the market today, they are all relatively similar. Whichever one you choose, I suggest that you install a coated polarizing or UV filter over the lens. It will reduce sun glare and as a bonus it will protect the lens from scratches and crashes.

A tip about these CMOS cameras: they all could suffer from "jello" effect, it's when the video wobbles around or when propellers look distorted. This is an issue with rolling shutter, because CMOS cameras capture by scanning the image in lines. CCD flight cameras are immune to jello because they capture the whole scene instantly. Quadcopters can jello more because they vibrate more than airplanes.

To get rid of jello: mount the camera on shock isolating pads such as Moongel, and remove all vibration from motors or props by balancing. If that doesn't get rid of jello, next reduce the shutter speed if you have manual settings or install a ND filter to trick the camera into a slower shutter. Next is to try and increase frame rate in the camera settings, 60p or 120p will appear smoother than 24p or 30p. (p = progressive frames per second). If all that fails, then mount a brushless gimbal. (http://amzn.to/1yI5J1b)

Camera Stabilization – When using a "gimbal" mount for your main camera it will remove some shake that the wind brings. Necessary to achieve professional smooth video, there are small brushless gimbals will give your smoother operation but may cost more than a servo operated gimbal. Checkout Tarot T-2D V2

Common micro HD cams used (5"+ prop quad recommended. If gimbaled then 8"+ prop quad required)

GoPro - Hero3 Black or Silver or White editions. 1080P cameras. My Silver keeps on working after several crashes, amazing! $200-350. 74 grams. (Protip: be sure to update the firmware and use PNY brand cards to prevent freeze ups) (http://amzn.to/1o4vmiM) (http://amzn.to/12hpEmk) (http://amzn.to/183JUsv)

GoPro – Hero4 Black, same size as a Hero3 but has 4k30 video which is four times the resolution of 1080P. Very good video quality. $499. 88 grams. Also a

Silver edition is available, for $399 but only does 1080P but has a built in LCD. 83 grams. And a new inexpensive basic model CHDHA-301 "Hero" which just came out for sale for $129, might be a great little cam….but it's hard to find info on it because it's named simply HERO…like the old gen 1 camera, that's why I put the full model number. (http://amzn.to/1vIlsYx) (http://amzn.to/1uSnxUQ) (http://amzn.to/Zabjcd)

Mobius action cam 1080p – a great little camera that is a third of the cost of others and is lighter and smaller, image quality is very good. $75. Only 39 grams, best for 220mm quads! (http://amzn.to/ZqJ0Hl)

Contour ROAM2 or the Contour+ 2, these are somewhat older and we some of the first "action" cams available. They are sturdy are work well. They have a new Contour 3 that just came out at $199. (http://amzn.to/ZCGtSB) (http://amzn.to/13rBo7m)

Sony HDR-AS10 or Sony HDR-AS15 or the new Sony AS100V, AS30V or the even newer HDR-AS20….they are all decent cameras and some have image stabilization! (http://amzn.to/ZAO19u) (http://amzn.to/11mjPR5) (http://amzn.to/1o4wkLR) (http://amzn.to/1p1SIFY) (http://amzn.to/1r0iRJy)

808 Keychain cam's – I can't recommend these because the image quality is not as good as the others, but they are very tiny and cheap. If you need tiny then I would get a Mobius unless your plane is so small it can't carry it.

…And with any of these cameras you are going to need some CLASS **10** or greater SD memory cards, class # refers to how fast the card can write; **10**MB/s (megabytes a second). A Class 6 might work for just taking snapshots, but for HD video a class 6 is too slow and may freeze your camera.

I have had some issues with some brands of SD cards in my GoPro 3 Silver, after a lot of frustration and trial and error I know that only this PNY 32 GB P-SDU32G10-AZ works well. (http://kiloohm.com/10/)

For other brands of cameras, these Sandisks may be okay as well as the PNY above:

SanDisk Ultra 64 GB microSDXC Class 10 (http://amzn.to/YyfCb3)

SanDisk Ultra 32 GB microSDHC Class 10 (http://amzn.to/13leUBj)

Note: Beware of counterfeit SD cards. Only purchase from reputable dealers and inspect the packaging. And perform a speed test of the card to see if it is indeed a class 10+. HDtune or CrystalDiskMark are programs that can benchmark your card.

46

Common "point and shoot" 1080p HD cams used (Gimbal required: 10"+ props quad to 15" hexacopter)

Sony Nex-5R – 16 megapixels, aps-c, $500

Sony Nex-6 – 16 megapixels, aps-c, $500

Panasonic GF3 – 12.1 megapixels, micro 4/3, $280

Panasonic GF5 – 12.1 megapixels, micro 4/3, $280

Panasonic GF6 – 16 megapixels, micro 4/3, $500

Panasonic GH4 – 16mp, micro 4/3

Gimbals used for this class of camera: RCTimer ASP, Foxtech Zenmuse, X-Cam, or homebuilt

A few of you out there will want to carry a big DSLR or Cinema camera. To do that you will need an octocopter or octo quad and a large gimbal, all of which is going to run you $2000-10,000 depending on your overall payload weight. Also, the camera gimbal should be steered by someone who is not flying, a cinematographer perhaps.

Common "DSLR" HD cams used (est Gimbal required: 14" prop quad, 10" prop octocopter)*

Canon 5D Mark II / III
Canon 7D
SONY α900
NIKON D900/D800E/D700/D800

Gimbals used for this class of camera: Eagle Eye BL, Tarot Invincible Rabbit, or homebuilt. (6.5lb loading weight)

Common "cinema" HD cams used (est Gimbal required: 15"+ props <u>octocopter or 1000+ class heli</u>)*

Red Epic / Scarlet

Sony FS700 / FS1000
Canon C300 / C500

Gimbals used for this class of camera: Align G800 gimbal, Cinestar, Porta Head, or homebuilt. (11lb loading weight)

*depends on lens and overall weight.

Flight Cam Video Transmitter, Onboard

Frequency choices for video:

900MHz – Can be nightmare in the city due to interference from cordless phones. But otherwise a good choice.

1.2GHz…Europe legal only, but a good choice if you can use it. An issue will be sourcing off the shelf antennas, not many exist and will probably have to be made in house. There are several tutorials online on antenna making.

1.3GHz (1258 and 1280MHz) - better than 900MHz, a nice quiet frequency for low interference. Goes around buildings and trees well. Don't use a 2.4GHz RC radio due to harmonics without a 1.2GHz filter on the RC system. Also use a filter on your RCtx if you are using a 433MHz LRS, or place the receiver a few feet away from your RCtx. Requires an easy to get HAM license. Please visit the FCC's Website for more information. (http://wireless.fcc.gov/services/amateur/licensing)

2.3GHz - Performs like 2.4GHz but without the interference from local WiFi. May require license.

2.4GHz – Lots of Wifi interference. Decent for mid-range flyers. 1-4 miles est line of sight…or through just a few rows of trees.

5.8GHz – Very common for inexpensive FPV. But only acceptable for a "park flier", because they don't have good range, ¼ mile max. But they do have tiny components which enable an installs into micro RC airplanes. Trees and buildings completely block signal. Should use circular polarized antennas to negate multipath reflections/glitching.

The next consideration is **Transmit Power**. The FCC only allows 36dBm EIRP total. EIRP is the total of the transmitter gain (dBm) and gain of the antenna (dBi), minus cable loss of which ours will be close to 0. Our antennas on our vTX will be omnidirectional with a gain of 0dBi, so our transmitter can technically be as high as 36dBm which equates to 4 watts of power. But really that is too much power in close proximity to the other electronics and with 4W it may cause issues with bleed over and servo glitching. 500mW (27dBm) or 800mW (29dBm) will both yield good results with no bleed over and be good for 5-40 miles line of sight depending on what ground station reception antenna is used. Another thing to consider is power usage, a big-ole 4W transmitter will eat your battery fast! So choose the lowest power that will work for you. Park fliers: 100mW is fine.

Be sure to mount the transmitter as far away as you can from any servos and GPS and RC antennas!

Video transmitter brands:

Lawmate has a good reputation.

Immersion RC has a good reputation. Only 5.8GHz is available right now.

SierraRC – The one I use, 800mW 1250&1280MHz. They are cheap and work. On the surface it looks like one of those no name Hong Kong generics because it sports a similar aluminum casing, but it has good components and a good track record. I remove the casing of the Vrx and add a small heat sink to reduce overall weight. They have other power options to choose from.

Tip: Avoid eBay generics, they are not calibrated!

My shabby 250mm mini-H quad. The silver box is a 800mW 1.3GHz video transmitter from SierraRC. I used some thermal transfer tape to help cool it. It's probably too much power for this little rig, the whole thing is the size of a soda can. You can also see the homemade Micarta blend composite for the top plate, 2 layers of old jeans and one Kevlar and one carbon.

Antenna Styles for Onboard Video

The common styles that are used for the **video** are the same styles as the RC system, but the frequencies and physical size will be different: **Whip/Monopole, Dipole, V-Dipole will work but a Skew Planar Wheel is preffered.** Note that these are all omni-directional antennas.

Next we will then categorize by their polarization, **linear (aka vertical) polarization** or **circular polarization.** A linear antenna is that old style antenna that is just a piece of straight wire usually called a Whip or Ducky, while it works, it can suffer from multi-path interference which is fine for RC but a problem for video. So if a wave bounces off a building it will arrive slightly before or after the direct wave hits the receiving antenna, this causes warping and lines in your video. To fix that we will use **circular polarization**, the antenna is physically twisted and so the wave projects out like a corkscrew. If the wave hits a wall it will rotate the other direction and cancel its self out at the receiver, thus eliminating multi-path interference and will provide crystal clear video.

NOTE 1: Circular polarized antennas come in either twisted LEFT or RIGHT. Be sure to use same-same for your transmitter and receiver, use either all left polarized (LHCP) or all right (RHCP) polarized. Do not mix together left and right polarized antennas, they won't work together. Most are right hand circular polarized (RHCP), so just buy or build all right hand. But if you are running multiple airplanes you could have your buddy use left hand and you use right hand to reduce some interference from each other.

NOTE 2: Linear and circular can be mixed, but there can be a slight drop in dB using a circular antenna in conjunction with a linear style. Example: using a Cloverleaf on the airplane and a Whip on the base station will result in not optimal (but still fair) situation. Just use all circular. Replace the Whip on the base station with a Cloverleaf or another style of RHCP antenna for best performance and clarity.

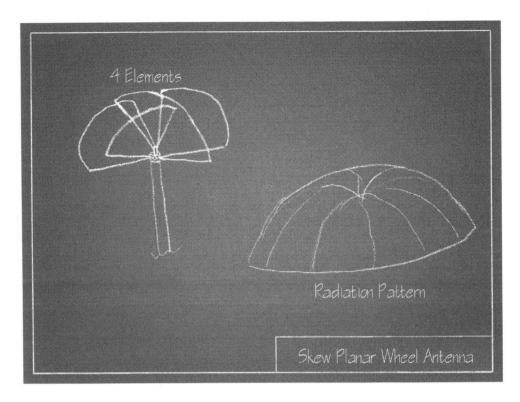

4 Elements

Radiation Pattern

Skew Planar Wheel Antenna

Skew Planar wheel – Omni, <u>circular polarized</u>. There are a few styles, each with more lobes/elements (wire loops). 2 elements = Butterfly, 3 elements = Cloverleaf, 4 elements = Skew Planar. 3-4 lobes are the most common and are the *preferred choice for video* because they provide a nice area of coverage and negate multipath interference. People are experimenting with more lobes but the more lobes that are added the less efficient they are at transmitting, but there may be something better in the future with more research.

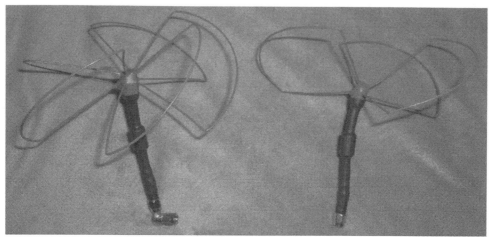

Video Aerial Systems "Ibcrazy" Blubeam Ultra 433MHz, receive on left, transmit on right. (http://amzn.to/1rSws57)

LHCP skew planar wheel that I made from Ibcrazy's design online, works good!

Chapter 5: Power System
Battery Types

Electric planes are not entirely a new thing, people have been flying electric since the 1980's, but they would only get a few minutes of flying in a circle then land quickly. What makes RC planes really cool now-days is that we can actually fly them some distance because we now have the juice and they outperform gas or nitro runtimes is most cases. Batteries are constantly being updated with new technology, but they still have their issues. Here's a list of some that's available. Most RC'ers use LiPo for all components.

LiPo – Most popular for RC. Very Lightweight. Overcharging can cause an explosion or fire. Very good energy density. Found in laptops and cellphones, also the current best choice for RC planes. Think about it, the tablet or your laptop are both stellar performers in the power department, they last days or weeks on a single charge and are super thin and lightweight. Consumer technology is why we are able to have an hour or two of flight time for our little hobby plane.

Lithium-Ion – Lightweight. Good energy density. Can explode in certain circumstances: submersion in water, short circuit, and overcharging. Commonly found in older cell phones, laptops and hybrid cars. Some quadcopter guys are in an unofficial internet competition and breaking records with <u>1.5 hour long hovering flights</u> with Panasonic NCR18650b in 6S with 26" props configuration. They have a very low C discharge rating, so low current slow-fly-big-prop setups are the only way to use lithium's.

LiFePoF4e – Lightweight. Safe. Large physical size for the power you get. Not readily available or popular due to poor energy density.

NiCad – Heavy. Reliable. Safe. Commonly found in power tools and commercial aircraft. Not suitable for small FPV due to weight.

Lead Acid – Not a good idea for FPV, too heavy. Commonly found under the hood of your car or Tactical-FPV-Van. Could be used for powering base stations. Be sure to use a SEALED type, else they will leak acid.

NiMh – Lightweight. Safe. Comparable to a NiCad but with more capacity (energy density). These are your <u>Sanyo Enloops rechargeables</u>. Good to use in radios and ancillary equipment like your FPV TV. (http://amzn.to/1168Niy)

Something new to think about:

Experimental Fuel Cell (Methanol to Direct Current) – Lightweight. High energy density; 8-10 times more than batteries. Safe as gasoline. Theoretically possible to get 8-16+ hours flight time. Easily storable in liquid form at ambient temperature. Methanol is commercially available everywhere at home improvement stores by the gallon or at race car shops.

Some laptop manufacturers are experimenting with this a long lasting battery augmentation, but not as a whole battery replacement. I believe this could be the next big thing for UAV's, a hybrid Methanol fuel cell with LiPo battery backup.

Experimental Fuel Cell (liquid or gaseous Hydrogen to Direct Current) – Lightweight, but storage container is heavy. Very High energy density; 700% more than methanol. Potentially explosive. Difficult to store, must be a pressurized vessel. Automobile manufacturers have experimented with this and have had good results technologically, however it is too difficult to transfer and store fuel and who wants to drive around a Hindenburg? But could be a future fuel for drones.

Various batteries, chargers and fire resistant bags. Note the bottom left 2.2 transmitter battery, it is puffed up with hydrogen gas and needs to be carefully recycled. That's what I get for leaving it fully charged in the radio.

Battery Sizes

When shopping for batteries there are 4 things to consider: voltage, capacity, discharge rating and quality/brand reputation.

Voltage (vdc)

Unlike other batteries that will list <u>voltage</u>, LiPo batteries use a nomenclature of how many cells are in the pack. **(4S)** would mean **4 cells x 3.7 volts =14.8 total volts.**

1S = 3.7 volt battery = 1 cell x 3.7vdc
2S = 7.4 volt battery = 2 cells x 3.7vdc
3S = 11.1 volt battery = 3 cells x 3.7vdc
4S = 14.8 volt battery = 4 cells x 3.7vdc
5S = 18.5 volt battery = 5 cells x 3.7vdc
6S = 22.2 volt battery = 6 cells x 3.7vdc
8S = 29.6 volt battery = 8 cells x 3.7vdc
10S = 37.0 volt battery = 10 cells x 3.7vdc
12S = 44.4 volt battery = 12 cells x 3.7vdc

Capacity (mAh)

Battery capacity, kind of like how big your fuel tank is.

Example: Your plane has a 1 amp motor. This means it uses 1 amps every second. So if I had a 1000 mAh battery.... then the battery would power the motor with 1000 mA for one hour.

Real world example: a quadcopter with motors that use 12A each at full throttle that we tested on a bench. Our quad hovers and flies at half throttle, so 6 amps times 4 motors. 24 amps est draw while hovering, aka 24,000mAh and I have a 1800mAh battery. I would need a 12,000mAh pack to fly for a half hour. I would need a 6,000mAh pack to fly for a 15min. I would need a 3,000mAh pack to fly for a 7.5min. At 1800mAh it should fly for about 4.5min.

Note: These numbers assume 100% efficiency and that's not possible. Also each pack weight affects the throttle position which alters the formula.

Discharge rating (C)

Discharge rating for batteries. Example; using a 2100mAh 10C battery,

(2.1x10) = 21 Amps

You can safely draw up to 21 Amps continuously without doing damage to your battery. Relative to motor sizing.

Higher C ratings are better batteries, and more expensive.

If your battery heats up while flying, you need a higher C rating. A little warmth is fine though.

You can do the math based on your motor draw to determine what C rating you need or use the RC calculators at *ecalc.ch*

Brand

A few good LiPo brands:

Turnigy Nanotech
Zippy
MaxAmps
GensAce
EP Buddy Glacier

….there may be other good brands, there is always something new out there.

Battery Connectors

This might not seem like a big deal at first but these connectors are the arteries of your electrical system. Beware of fakes out there, the wont handle the rated amperage or come loose.

<u>XT60</u> Hextronic - Handles 60 amps continuous, my favorite because they hold tight and just work. I also use XT60 for speed control to motor connectors. (http://amzn.to/1rSIThe)

<u>XT90</u> Hextronic - Handles 90 amps continuous. Good for giant builds. (http://amzn.to/1xY6mT8)

<u>Deans T</u> connector – An old standard connector but a problem is that the plastic casing melts when soldering, and become misaligned. To fix that: plug another Dean T into the other end to hold it together when soldering. The other problem is that there is no published amperage rating that I could find! Some guys say 30-60 amps continuous. For that reason alone, I won't use them. (http://amzn.to/1vKjV6d)

<u>Anderson Powerpole</u> – A fine connector. Crimp on, so reduced stress due to no soldering. Comes in various amperage ratings from 55 to 350 amps. Just be sure to use a proper crimp tool. Hard to find already attached to batteries. (http://amzn.to/1s68Coa)

<u>E-Flight EC3</u> – A good connector but again hard to find batteries with it already installed. 60 amps (http://amzn.to/ZPrZXO)

<u>E-Flight EC5</u> –some EC5's come with 4mm and some with 5mm plugs which could be a mess and waste of time and money is mixed up. 150 amps continuous. (http://amzn.to/1t0laza)

Bullet / banana connectors – Generic junk stuff. There are better options. If you must use it, be sure to lock them in place with shrink tubing.

Soldering a XT60

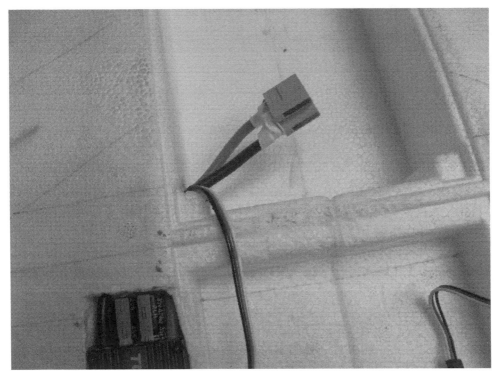

Dual battery parallel mod for longer flight times. Two XT60's and some solid core. And its lighter, no extra wires. Note there is a sliver of credit card plastic CA glued between the two to space them apart.

Wire Gauges

This probably won't be an issue for most of you to worry about because each component (esc's, batteries) comes pre-wired with the proper gauge leads, but sometimes the manufactures get it wrong because they guess as to what your current draw will be.

First determine the max amps your motors and gimbal and all equipment can draw. Then use this chart to check that the wire size that came with your battery is correct.

Estimated wire sizing for the chart below; <u>.5 meter length of copper multi-stranded wire</u>, 2% loss:

Battery voltage, Current draw, Wire size
1S 2.5A draw = 18 awg
1S 5A draw = 15 awg
1S 10A draw = 12 awg

1S 15A draw = 11 awg
1S 20A draw = 9 awg
1S 25A draw = 8 awg

2S 2.5A draw = 21 awg
2S 5A draw = 18 awg
2S 10A draw = 15 awg
2S 15A draw = 14 awg
2S 20A draw = 12 awg
2S 30A draw = 11 awg
2S 40A draw = 9 awg
2S 50A draw = 8 awg

3S 2.5A draw = 23 awg
3S 5A draw = 20 awg
3S 10A draw = 17 awg
3S 15A draw = 15 awg
3S 20A draw = 14 awg
3S 30A draw = 12 awg
3S 40A draw = 11 awg
3S 50A draw = 10 awg
3S 60A draw = 9 awg
3S 80A draw = 8 awg

4S 2.5A draw = 24 awg
4S 5A draw = 21 awg
4S 10A draw = 18 awg
4S 15A draw = 17 awg
4S 20A draw = 15 awg
4S 30A draw = 14 awg
4S 40A draw = 12 awg
4S 50A draw = 11 awg
4S 70A draw = 10 awg
4S 85A draw = 9 awg
4S 100A draw = 8 awg

6S 2.5A draw = 26 awg
6S 5A draw = 23 awg
6S 10A draw = 20 awg
6S 15A draw = 18 awg
6S 20A draw = 17 awg
6S 30A draw = 15 awg

6S 40A draw = 14 awg
6S 50A draw = 13 awg
6S 70A draw = 12 awg
6S 90A draw = 11 awg
6S 110A draw = 10 awg
6S 120A draw = 9 awg
6S 150A draw = 8 awg

If you need more or want to adjust the cable length for voltage drop then do a google search for "DC Cable Sizing Calculator", lots of solar panel companies have these online calculators.

Battery Chargers

Danger: Charge batteries inside fireproof box/bag or on an open concrete floor. Lithium and LiPo batteries may explode or catch fire if overcharged or physically damaged or submerged in water.

People's homes have been destroyed because of negligent use of LiPo batteries!

-Don't overcharge.
-Don't leave unattended when charging.
-Store in a fireproof bag/box.
-Don't store them with a full charge.
-Keep a fire extinguisher nearby.
-Use a quality charger in the correct mode. (Voltage, capacity, battery chemistry.)
-If the battery is puffed up, don't charge it. Safely recycle it. It may burst.

Good Chargers
<u>Turnigy Accucell 6</u> ...requires a 6A 12V power source: computer power supply or car battery. A good charger none the less. One weird thing is you have to hit start several times to get it to start charging. $20 (http://bit.ly/1vRxbUQ)

<u>B6AC</u> – Essentially an Accucell 6 with the same display and user interface but with a built in 120VAC wall and car power supply and passively cooled. Excellent charger. $35 on Amazon Prime. (http://amzn.to/ZrnfH4)

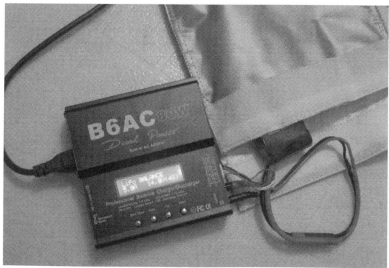

<u>B6AC</u>

Chapter 6: Ground Control Station

The base station/ground control is just the other side of the Radio Control and Video Systems. In this chapter we will discuss how to build a quality base station, transmitters, different ground based antenna types and mapping integration, and a bunch of other cool stuff!

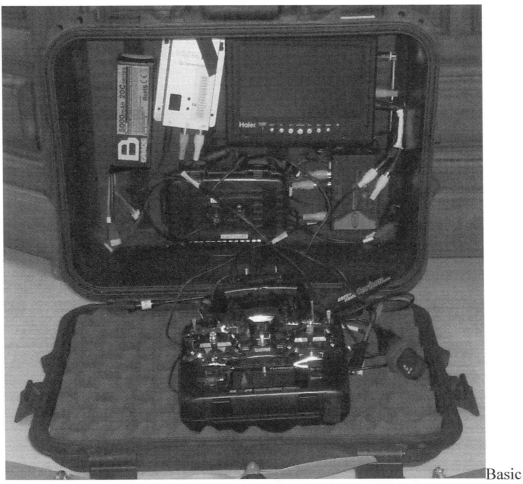

Basic FPV/UAS setup. 5000mah 3S battery, 7" LCD, DVR, Powered video splitter, RTL sdr, video rx, fuse box, 9XR. (not shown: laptop with mission planner and USB video recorder).

Handheld RC Transmitter

We technically need only 3 channels for a basic flying wing (left and right elevon and speed control), but once you start adding things like autopilot, RTH functions or a rudder or gimbal they will each use a channel, so just get the most channels you can afford and is compatible with your LRS. 3 to 6 channels is typical for airplanes and quads. 6-8+ Channels is typical for heli's.

Take note that some of these feature a modular removable back piece module to swap out frequencies and power easily; Long Range System 433MHz (LRS). Alternatively some LRS's use a cable that plugs into the "trainer port" of a transmitter, or worst case would require soldering. Some LRSs don't work well with some transmitters, so if you can- choose your LRS *before* you buy the transmitter.

The *sticks* will come in "Mode 1" or "Mode 2", which means the throttle stick is located on the left side for mode 2 and on the right for mode 1. Typically mode 2 is used for airplanes and multicopters. Mode 1 is for Heli's. But there is no reason you couldn't use either one for whatever you want.

Turnigy 9XR, sheet of paper is the APM switch modes

Popular RC transmitter models that work well

Flysky TH9x, $100 (http://amzn.to/1siCT47)
Turnigy 9X, $50 (clone of the TH9x). (http://bit.ly/1BXBRdE)
Turnigy 9XR, $59 (9X bigger brother with better ER9X firmware. But no radio module included.) (http://bit.ly/1uGR5Ue)
Turnigy 9XRpro, $99 (9XR bigger brother with built in Arduino, telemetry, voice and a SD card. But no radio module included.) (http://bit.ly/1o1HmXY)
FrSky Taranis $200 (Has OpenTX firmware, similar to ER9X. Telemetry downlink capable) (http://amzn.to/1tGvas0)
Spektrum DX6i, $125 (http://amzn.to/1slrPCm)
Spektrum DX7s, $250-300. (http://amzn.to/1t5hvAn)
Spektrum DX8, $300. (http://amzn.to/1t5hmga)
Graupner MZ-24, $500. (http://amzn.to/1o1GpyP)
Futaba 10CG, $650, (http://amzn.to/10W3wjH)

Below is a diagram of the radio settings of a Turnigy 9XR for use with APM/Pixhawk. Could also be used with any radio with ER9X or OpenTX firmware such as the FlySky Taranis or Turnigy 9X or Flysky TH9X or 9XRpro. The menu's may differ slightly on other radios.

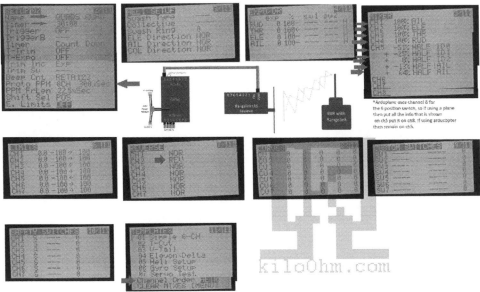

Large image link (http://kiloohm.com/?p=388)

I'm using a APM 2.5 with Arducopter firmware on a quadcopter. If you or using a gimbal the remaining radio channels can be used for that. Ch5 is for the 6 position switch in ArduCopter. If using ArduPLANE, that switch moves to channel 8!

CH2 may be reversed or not depending how you like the elevator. It may be revered by selecting reverse on screen 7, or -100 on screen 5.

In mission planner set my 6 position switch to: Stabilize, Auto, Althold, loiter, land, RTL.

Important! For the 6 position switch (ch5 'copter or ch8 'plane) your values may vary slightly. My numbers might get you close but you should double check in Mission Planner. Every radio performs slightly different even if it's the same model.

Also featured is a wiring diagram for PPM. PPM is great for reducing the amount of wiring. Rangelink, EzUHF, OpenLRS are all good PPM capable 433MHz 1W systems. If you are unsure if your receiver is PPM capable then try the DSM2 selection on menu 1 of the 9XR and add a wire from each channel of the receiver (1,2,3,4,5 and so on) to the APM inputs.

Antenna Tower/Tripod

You can use anything that will hold your antennas steady, a stick or pvc pipe, whatever. Preferably it will be at least 6' high so it "sees" above people and fences. I like some portability so I use a cheap collapsible monopod that I mount on top of a tripod. I had to make an adapter which is just two ¼-20 nuts welded together end to end to make a long nut with some spacers added on the edges. The overall height of the mono pod is variable between 2' to 6', then you attach that onto your camera tripod you can get to 10 or 15' overall depending on your tripod model!

15" antenna with LCD monitor

All folded up

For more info on how to construct this tower:
http://kiloohm.com/the-perfect-fpv-uav-radio-antenna-tower/

Antenna Styles for the Base Station Video

The antennas are some of the most critical of components, more than transmitter selection. Increasing the dBi of your antenna will give you more total power (EIRP), and cheaper than just using a more powerful transmitter (dBm). The reason is because of efficiency of design, gain. A 3dBi antenna is four times as efficient as a 1dBi antenna and you need 6dBi (4X the power) to go twice as far. To increase your dBi, you can ditch your omni use a directional antenna like a patch or Yagi, these will get you up to 10-25dBi respectively. If you want super range, you could use a dish which would grant you 40dBi and get you into outer space but it would be extremely difficult to aim because its beam is so narrow.

Base station antennas will be first categorized into **directional** and **all-directional**, AKA uni-directional and omni-directional respectively. *Uni-*Directional or just *Directional* means that the antenna 'projects' out like a flashlight, they will have less interference and will work better at long distances. Omni-Directional works in all directions, but will also receive interference from all directions.

Here's an overly simple way to think about how an antenna can be made to be more powerful (in one direction); Imagine a candle burning in a dark room, this would be the equivalent of 0dBi "omni directional" stick antenna, all the light is shining in all directions. But then you put a reflective face mirror behind it and now all the light is shining on only half the room (180 degrees) and seems a bit brighter on that side of the room and dark on the other, this would be like your "patch" antenna. Then you get the idea to remove the mirror and replace it with a shiny metal salad bowl, this is your "dish" antenna (364 degrees approximately). The room is mostly dark except for where the bowl has focused it to a tiny point on the other side of the room.

Now unfortunately, directional antennas are too large and require aiming so they can't be used on our FPV airplane, you would need a plane the size of a large car to fit a directional antenna and its aiming system. So, all the antennas on the FPV plane will have to be Omni directional. On the base station we will always use a directional antenna, except for park flyers where they are playing around overhead in many directions. If you are park flying then use any of the omni antennas listed in the RC Chapter 3 for your base station as well as your airplane, just be sure to select to appropriate frequency for your Video or RC system.

Before proceeding be sure to read about the onboard video antenna descriptions in: Chapter 4.

The following are the preferred long range <u>directional</u> styles that are used for the **Base Station**: *(antenna styles for Video and RC are covered in their respective chapters)*

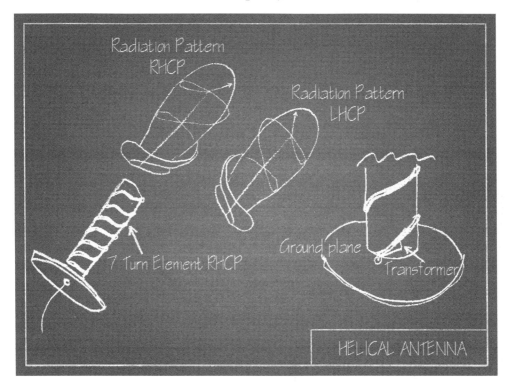

Helical – 7-20dBi gain, *Circularly* polarized. Capable of 30-40 miles+ est. with 500-800mw. Available in 3+ turns in length, adding more turns gives more directional gain at the sacrifice of some wide beam. Here is a link to an ibcazy 2.4Ghz helical. (http://amzn.to/1uu1bYu)

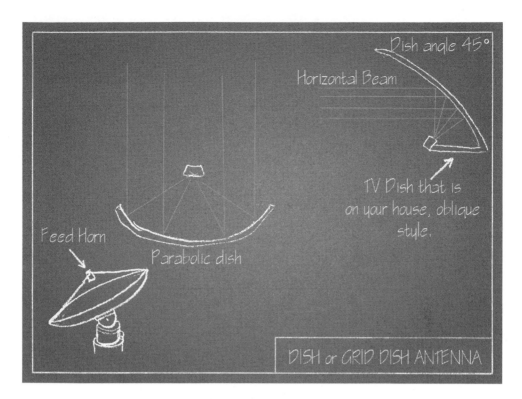

Parabolic Dish – 20-50dBi gain, Reach into outer space with one of these. Repurpose the one on your house. Linear or Circular polarized depending on what feed horn is chosen. Requires a complex GPS guided automatic aiming system. Use in conjunction with other antennas for reliability. Provides the most gain but also the most hassle.

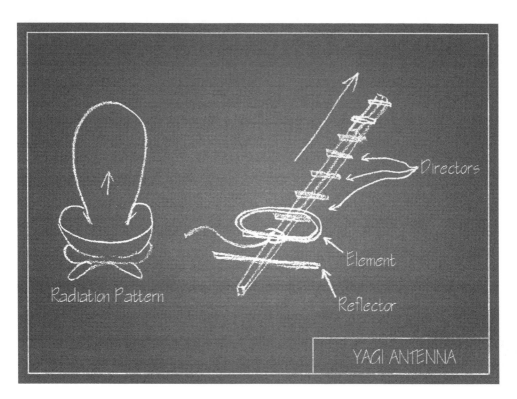

Yagi – 13-20dBi gain, Linear polarized. Commonly seen on rooftops of houses as large TV antennas, but for our frequencies they would be physically smaller about 22" long for 1.3GHz. If possible choose a circularly polarized Helical for reduced multi-path interference. Capable of 30-45 miles+ est. with 500-800mw.

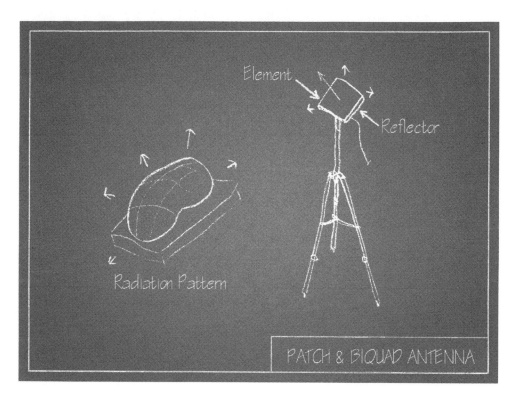

Patch and Bi-Quad – 8-12dBi gain, Linear polarized. Both are different but similar in function. If possible choose a circularly polarized antenna for reduced multi-path interference. Capable of 5 miles+ est. with 500-800mw. Small and portable.

<u>Crosshair</u> / **Pepperbox** – 10dBi~ gain. A new antenna design that features <u>circular</u> polarization but in a small package the size of a bi-quad or patch, in direct competition with helical antennas. I'm unsure of range, but I would expect it to be around 5 miles. (http://amzn.to/1nb5iHH)

Additionally if you don't want to lug around a large directional antenna and just plan on flying short range you can use a small omni. The common omni short range styles that are used for the **onboard RC and video**: **Whip/Monopole, Dipole, V-Dipole, Skew Planar Wheel.** Note that these are all omni-directional antennas and should only be used for short range park fliers!

Video Receiver/Tuner

This section is about the interface box (receiver) that fits between your video antenna and your TV or goggles.

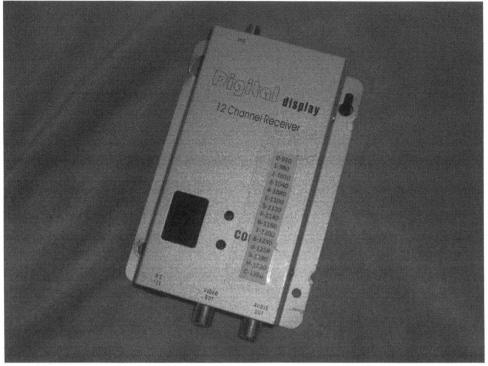

The best unit I've found: SierraRC receiver with **upgraded SAW filter and Comtech tuner**. Note the Channels/freq chart that it can receive. Does everything except 2.4GHz and 5.8GHz. Requires 12V.

Brands: RMRC, Lawmate, SierraRC, Rangevideo, (avoid generics!).

Planning on using a multiple monitors? Got a passive video splitter or will just twist the wires together? Be careful, it will reduce your video quality. Use a powered splitter instead!

Powered Video Splitter / Video Distribution Amplifier

A powered video splitter just takes one source of video and sends that to multiple outputs. Don't try and splice three composite cables together, it will mess with your ohms and give you a reduced video quality which will make you crash, but that won't happen with a real deal **powered** video distribution amplifier. Most use

12vdc. There are many places to get them; RadioShack, Amazon, car stereo installation shops for mounting TV's into auto headrests, etc. (http://amzn.to/12B2qWP)

Bonus tip: Diversity boxes and Antenna trackers sometimes have powered video splitters built into them!

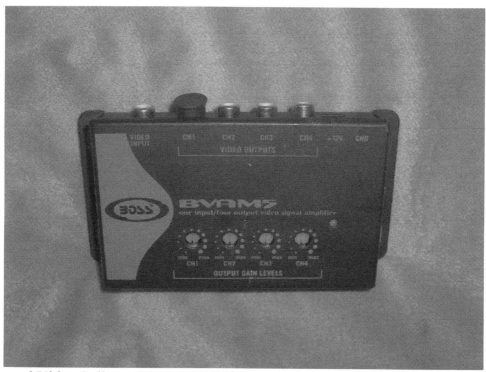

Powered Video Splitter, BOSS BVAM5. Requires 12V. (http://amzn.to/1vJGSUX)

LCD TV

A single LCD screen can be your entire monitoring setup, or as a secondary monitor if you have friends with you. It gives them something to look at if you are wearing video goggles.

Beware though that so often with TV's they have a "feature" that when static is on the screen it will cover it up by splashing the screen with blue so you don't have to look at the horrible static, I guess the manufacturers think blue is better but when choosing a TV for FPV, choose one that does <u>not</u> "go to blue". Static happens while flying and it's the last thing you want is your display to go out to the blue screen of death. At least with a little static you can continue flying.

The <u>Haire 7" portable HDTV</u> features good battery life and works well in open sunlight, and it doesn't go blue on static. I got mine refurbished on eBay for about $40 and it works great. I have also heard good things about the <u>Digital Prism</u> and Phillips 7012 & 9012. Or Foxtech has a 7" with a built in 5.8 receiver and DVR, way cool! (http://amzn.to/Z6YAEM) (http://amzn.to/13BFkjQ)

Your monitor doesn't have to be just a 7" screen. Some guys use 50" LED HDTV's in the back of their vans, sort of like a mobile command center. Really cool, however it would require you to own a van. Most everyone else opts out of the van and goes for the head mounted video goggles.

My portable FPV setup. Haire 7" LCD, Coroplast sunshield, Turnigy 9XR with OpenLRS, <u>Ibcrazy Blubeam 1.3 antenna</u>, Sierra 1.3 receiver…and scrap aluminum license plate bracket as a frame. All runs off a 5000mAh 3S. It's so ugly, but works so well! (http://amzn.to/1rSws57)

Video Goggles/Headset/Video Glasses

Video glasses basic requirements: 640x480 or greater resolution. No blue screen. Diopter if you use corrective lenses, otherwise not required.

Fat Shark – The standard FPV glasses. They are the standard for a reason, they are purpose built for FPV and work. Fatshark is an ImmersionRC company which has a great rep. The Dominators, Predators and the Base are all good. Diopter lenses available separately. 640x480 resolution. $250-350

Fat Shark Dominator V2 – A new 640x480 resolution headset with built in DVR and interchangeable RF modules. $380

Fat Shark Dominator HD – A new 800x600 resolution headset with built in DVR and interchangeable RF modules. $550

Vuzix Wrap 1200 or 920 – Semi-Blue screen's upon static, statics a bit then blue screens. Requires a ski goggle mod to block sunlight. Has diopter. 852x480 and 640x480 respectively. $400 & $350

Zeiss Cinemizer Plus – Requires a ski goggle mod to block sunlight. 640 x 480 pixel per eye. Has diopter. No blue screen. Discontinued. $200-300 eBay.

Zeiss Cinemizer OLED – Expensive. Requires a ski goggle mod to block sunlight. 870 x 500 per eye. Has diopter. No blue screen. Amazon $750 (http://amzn.to/1oKMnPg)

Headplay – Features 800x600 resolution in each eye which is one of the highest available, but it's a moot point because our SD cameras are only capable of 640x480. The Headplay has a "Liberator" box, which moves most of the electronics off the head unit, but it can be a boat anchor too. Has diopter. No blue screen. Headplay is out of business…so no warranty or support. $320est

MyVu Crystal - Inexpensive good quality 640x480 resolution glasses that can be snatched up for $100 used on eBay. But difficult mods are required to fix blue screen and white haze and dead batteries. I love these Crystals because they are cheap, tiny and lightweight and have a sharp screen. But no diopter adjustment. Requires a ski goggle mod to block sunlight. Do not get the MyVu "personal / solo / 1003i or Shades" models, they are low resolution. Also, MyVu is out of business…so no warranty or support. When they went out of business Google bought the patents and created Google Glass, its good tech. Don't forget to get a compatible RCA cable set.

There is also a rare EV model that has a 30% wider field of view, it's the one to get if you can find one. (http://kiloohm.com/?p=118)

New **Spektrum VS1100 Ultra Micro FPV System** - Co-developed with Fatshark & Spektrum, QVGA 320 x 240 LCD displays, 5.8GHz wireless receiver, RHCP antennas, Includes mini-camera. $350 Lower end resolution, but easy to slap together. Has something called "digital head tracking"? I'm not sure what that is…digital...hmm, we will have to wait until November 2014 to find out. Similar to the Fatshark Teleporter kit.

Laptop and Software

A laptop isn't something everyone needs with them in the field but it can give you some nice features like; mapping, missions, telemetry, spectrum analyzer and a disk video recorder. You won't need anything too special for computer hardware, just as long as it's capable of running Google Earth at a minimum and the battery holds for as long as you need it.

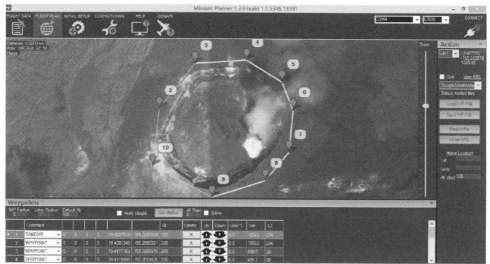

APM / Pixhawk Mission Planner

Mapping Applications

Ardupilot APM "MissionPlanner" – Best with 3D Robotics two way telemetry but also pseudo-pairs with XBee 2.4GHz or XRF 915MHz homebrew after a lot of tweaking, just get the 3DR telemetry. See QGroundControl.

MissionPlanner 2 / APM Planner 2 – Newest 3D Robotics planner based on Qgroundcontrol.

DroidPlanner – App for Android, used with Ardupilot APM hardware and 3D Robotics USB telemetry hardware. Is a nice way to take your mission planner to the field without carrying your laptop. Will require an Android tablet or phone, micro-usb OTG adapter and a 3DR telemetry kit. (http://amzn.to/1r2lw4r)

DroidPlanner 2 *beta* – Same as above but features a "Follow Me" feature. With this loaded on your phone it takes the phone GPS and feeds it thru the OTG adapter to the 3DR radio and to the APM or Pixhawk. This would be great for following a water skier or runner. It would be like having a pet follow you around.

Openpilot – Ground Control Station. For use with…you guessed it!

uThere Ground Station – pairs with RubyOSD.

ImmersionRC – Uses telemetry and plots on Google Earth. Pairs with either ezOSD or TinyTelemetry. Not a full ground station solution.

DJI NAZA Ground station – For Quadcopters only.

UgCS 2.0 – Yet to be released, universal ground control station that looks neat. "We support not just DJI (phantom2), but numerous other autopilots like: 3DR APM/Pixhawk, ArDrone, PixHawk, Mikrokopter, DJI autopilots, UNAV, Microdrones." –UGCS.

Telemetry Systems

Used for steering antenna trackers, mapping data, viewing live sensor data and locating missing planes by their last known GPS data.

One way down modem telemetry – Some OSD's will feed telemetry through the right or left audio channel of your Vtx system; ezOSD and RubyOSD. To use this, simply plug the cable into your laptop earphone jack to your Vrx audio out. Or with ezOSD or TinyTelemetry you can use your iphone or android with the app iTelemetry or on the PC with ImmersionRC software.

Other OSD's might use the closed captioning feature (invisibly) to feed the telemetry across; Eagletree EagleEyes.

Not all OSD's have telemetry features.

> If you want *two way modem* communication use one of these below, but it will cost you small amount in power and weight:

3D Robotics – two way/up down at either 433 or 900MHz, but only 100mW power. 1-2 miles range. USB connection. Use with Ardupilot and also works on other autopilots because the radios are just wireless serial connections.

uThere Telemetry - two way/up down 900MHz telemetry at 250mW and a 750mW versions. Ability to control Ruby Autopilot with this; upload waypoints etc. Only used with uThere Ruby autopilot.

These two below have 2 way telemetry built into their receivers.

FrSky – 2.4GHz, 60mW. Built into their receivers. Low power, no use for a long range FPV.

OpenLRS – Their 433MHz receivers have a 100mW transmitters built in for telemetry. But if you want 1000mW, you can use two of their RC transmitters and modify the firmware to put one of them in receiver mode.

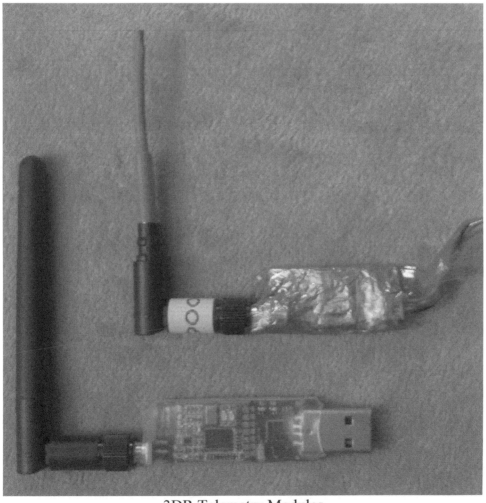

3DR Telemetry Modules

Digital Video Recorder (DVR)

There are several if not hundreds of these devices to choose from. Most nowadays are the size of a deck of cards and take the video feed and store it straight to a SD card and won't cost more than a $50 + an SD card. This is an easy way to record, but you won't have playback features or a monitor to view the video. I prefer using a computer and a capture device as described below for more control. *requires a powered 12V Video Splitter / Video Distribution Amplifier. (http://amzn.to/12B2qWP)

Easy to use models:
Aiptek mini-camcorder
Japan-FPV DVR

Disk Based Video Recorder (laptop)

Instead of a DVR you can record video straight to your laptop. You will need a video capture device, there are several to choose from here on Amazon for $20-$40*. I have yet to find a perfect one yet so I won't recommend any specific one, they all have their idiosyncrasies so please read the reviews to see what works best for your system. I use the EasyCap with Windows 7. You will need a fast CPU. (http://amzn.to/14dijTW)

Then you might need some software to capture, you can could use a great program called FPV Recorder, or VirtualDub. I recommend using H.264, MP4, XVID or as your encoding codec. These video files will be large in size so you will need a large internal hard drive in your laptop, or you may use a USB-3.0 external drive as long as your laptop supports 3.0. Don't use an external drive if your laptop only supports USB-2.0 because its IO bus will already be overtaxed with incoming capture video, there won't be any room on 2.0 for outgoing video. *requires a powered 12V Video Splitter / Video Distribution Amplifier. (http://amzn.to/18QPz6d) (http://amzn.to/12B2qWP)

My FPV recorder software settings for PZ020 camera and Easycap:
OEM Device
720x480
YUY2
composite
NTSC M
x264vfw (separately downloaded codec) or use MPEG for slower PC's
Mpeg layer 3 audio

Spectrum Analyzer

A $20 SDR is a nice piece of kit to add to your base station laptop. It's a receiver and spectrum analyzer. It will allow you to see with your own two eyes on your computer screen what is going on in your radio frequency environment. Use it for finding interference before you fly. You can also use it as an ADS-B receiver! (real time commercial aircraft telemetry tracker).

RTL SDR Hardware – An "RTL" is low cost to get into because they are re-purposed USB TV Tuners, retails for $10-$20. Plugs into your laptop with a small USB dongle and antenna. These RTL's only receive, no transmit. Right now they come in two varieties. Search EBay or Amazon for USB DVB-T, *RTL2832U* **R820T** or *RTL2832U* **E4000**. You won't find these on store shelves in the US, because they are originally for foreign TV reception, you can however buy them online. These cheap RTL dongles are perfect for monitoring a RC 433MHz LRS and 900MHz or 1258-1280MHz (1.3GHz) video signals!

R820T chipset. Frequency range is 24 – 1850 MHz, with no gaps. Newest available. $10-20. (http://amzn.to/17t9u7S)

E4000 chipset. Frequency range is 52 – 1700 MHz, with a gap from 1100 MHz to 1250 MHz. Some might go up to 2200 MHz. Do not use an E4000 due to gaps, might not work on your video channel.

Whichever one you decide on buying, it will need an upgraded antenna because the tiny TV antenna that it comes with isn't very good.

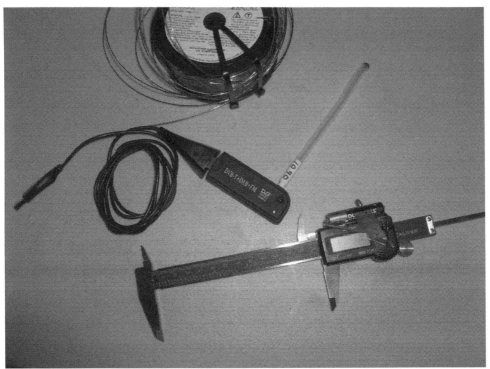

RTL2832U R820T with a homemade 1090MHz (ADSB) antenna from mig wire 131.2mm and a plastic tube. I have received and tracked planes over 200 miles away and two mountain ranges while indoors with this!

SDR Antenna – You will need to buy or build an antenna. And locate the antenna away from any laptops or other electronics, they all make noise. Look up: *Discone antenna* – Omni, covers a lot of frequencies, easy to build. Or you can just us a regular VHF/UHF television rabbit ear which is what I use. Additionally you can get an up-converter so that you can monitor higher microwave frequencies like 2.4GHz, but can be expensive and unnecessary for long range FPV.

SDR Software – Choose one of these software's to use with your RTL SDR or other SDR.

> **SDR# (SDR Sharp)** – For Windows 8, 7 and XP. Easy to use and setup. Quick start guide and nice wiki at: http://rtlsdr.org/softwarewindows or http://sdrsharp.com/

> **HDSDR** – Linux and Mac. Not easy to setup with an RTL.

> **GNU Radio** – Linux. Very powerful, has many software filters and processing capabilities.

86

Other Spectrum Analyzers / SDR Hardware – Costs $150 to $6000, but also may have transmit capability and cover more frequencies. Examples: *HackRF, BladeRF, Winradio, FlexRadio, RF Explorer 3g, BK Precision*

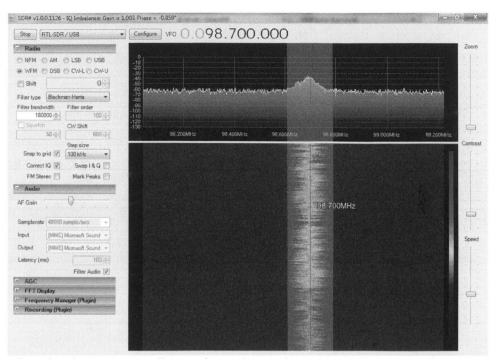

SDR# tuning into a FM radio music station, using a RTL2832U R820T with stock junk antenna.

Base Station Power

There are a many ways to power base stations. You can use individual LiPo batteries to power the TV receivers and the portable TV's, but that can get expensive buying individual LiPo batteries for each component.

An economical way to power a portable base station are the sealed non-spillable type, such as an Optima (75Ah) blue or yellow top. (http://amzn.to/Z2g5qu)

If you don't feel like wiring anything up and want plug and play 110V AC for laptops; this all-in-one inverter Duracell Powersource (60Ah AGM) will do. (http://amzn.to/11mkGkT)

You may already own one of those car battery jump packs, they will work for short periods, like this Duracell Powerpack 600 features 110V AC and 12V DC. (http://amzn.to/11iHVO4)

But if you want lasting power then a gas genny, like the quiet Honda EU2000i or Yamaha EF2000iS. You could charge your LiPo's in the field! (http://amzn.to/XGHgbI)

Automobile jump pack. This one has a 12V 18 ampere-hour battery. Works fine with a lipo charger.

Chapter 7: Drive Train System

Two things you probably would like to know- how much weight can it carry and how long can it fly? To figure this out there is an easy way and a hard way. To calculate the motor and prop matching and overall weight of the quadcopter; the easy way is to use the Ecalc.ch calculators to virtually build your whole system. From there it will give you estimated flight times and payload capacity. It's pretty good, but it's only about 80-90 % accurate and that's if you manage to fill out the diabolical form correctly. The hard way, which is the more accurate way is described below:

-Measure motor kv. Unloaded voltage test at 1V. Mark RPM.
-Propagate the chart below by "dyno'ing" the motor with several prop and voltage combinations through the full range of throttle. Measure: thrust in grams, current draw, g/W (grams per watt) (I like Soma's method: http://vimeo.com/87633657)
-if an airplane, wind tunnel testing. Measure lift and drag.

After dyno'ing motor/prop/voltages, you will have a data set that looks like the one below. Reputable manufacturers will already have this done for you. (thrust is in grams)

Example 780kv motor from brand X								
Volts	Prop	Throttle %	Amps	Watts	Thrust	RPM	g/W	Temp in C
11.1	1137	50	2.3	25	260	4300	10.4	40
11.1	1137	100	9.4	102	750	7100	7.35	40
11.1	1240	50	2.6	30	320	4000	10.67	44
11.1	1240	100	12	133	990	6700	7.44	44
14.8	930	50	2.7	40	300	6800	7.31	41
14.8	930	100	8	119	680	10000	5.71	41
14.8	1033	50	3.1	45	360	6000	8	42
14.8	1033	100	11	135	990	9500	6.05	42
14.8	1137	50	6.5	51	430	5500	8.43	45
14.8	1137	100	14.9	216	1260	9000	5.83	45

The 1240 prop with 3S battery has the best g/W (grams per Watt), it's the best for this motor. And would require a 15A ESC, that's providing 20% headroom. (thrust is in grams)

Let's say I want to carry a gallon of water. From here you can do some quick math. With that prop and battery we chose above; each combo will give us 320g at half throttle. Half throttle on a multicopter is generally the hover point we want. A gallon of water weighs 3782g and will take 11 of those motors to fly that gallon jug around, but that's not calculating our frame/battery weight, which is another simple addition. But we'll move on, since 11 motors is too many. This motor isn't the best for carrying a heavy load and would be better suited carrying a small payload like a GoPro. A GoPro with gimbal is 200g, our frame is 400g, and battery is 150g totaling 750g. So it will hover with only 2.34 of those motors, a quadcopter with 4 of those motors and props will fly nicely and could even work as a tricopter.

But how long will it fly my GoPro around at 50% throttle? A 3S lipo that weighs 150g turns out to be a 1800mAh pack of which we can only use 85% of due to the voltage gets lower near the end of the flight and also a small margin of error and safety. Which gives us 1540mAh of usable battery which we convert amperes per minute* = 92.4Ah.mn According to the chart, the motor will draw 2.6A at half throttle but we have four motors, so 10.4A total. Divide 92.4 by 10.4 and we get 8.88 minutes of flight at half throttle. This is very general estimate that will depend on weather and speed and other variables.

* 1 Ah = 60 A.mn

Motor

Brushless – low maintenance pulsed DC motors.

Size – A motor from company X has a motor called the XM3520/14, the XM means nothing, but the 3520 indicates the stator diameter (35) and stator length (20). The 14 indicates how many poles there are. Generally but not always it's the rule, sometimes it means nothing depending on manufacturer.

Outrunner/Inrunner – All the pictures of motors in this book are "outrunners" meaning the outer casing is the spinning part. An "inrunner" the outer casing is stationary and a shaft spins, like the old DC brushed motors. Inrunners are not optimal due to less cooling and more weight.

Bearings – VXB USA bearings are real good. NSK Japanese bearings are good. Chinese are not quite as good. But even better than all those stainless bearings is Boca ceramic bearings because they are quiet are super smooth, but they are very expensive. They all use the ABEC standard but some countries have been importing fake ABEC-7's when they are actually just ABEC-1, shop wisely.

Whatever bearing you use, it will require maintenance regardless if they are "maintenance free" they will need a drop of small bearing oil every 10 hours of flight. 10 hours just a guestimate, but it works for me. Each motor has two bearings, so don't forget the bottom one.

Popular motor manufacturers: T-Motor, Hacker, RCtimer, Hobbyking, Sunnysky, Cobra

$20 dead motor, notice the burnt windings. I flew it after a crash and it shorted out from a little bit of mud inside. The actual motor size is no wider than a quarter. The bearings are removed.

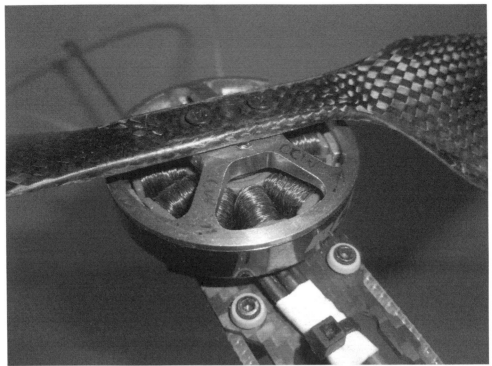

Brushless "pancake" style motor, RCTimer 360kv.

NTM 1250kv outrunner motor, APC prop. Gorilla glue on foam is repaired cracks, this thing has crashed a lot! Haha

Speed Control (ESC)

To size your ESC, you will need to find out how much amps your motor and prop combo will be using. Then take that number and <u>add 20% headroom</u>. If your motor uses 12A at full throttle like the example in the motor section, do not get a 12A ESC. Get a 15A to 20A ESC to allow for peak amps and it will reduce overheating problems with an over tasked and under cooled mosfet.

BEC- Battery Elimination Circuit. It outputs 5V that is used by the radio system. There are <u>standalone BEC</u>'s and some ESC's have a built in BEC. Clear as mud? (http://amzn.to/1yIra2l)

OPTO - is an optical coupler for your data lines, it is there to prevent back feeding high voltage to your radio and autopilot upon ESC meltdown. Opto ESC's are a good thing and cheap insurance for just a few bucks.

Voltage: Be sure to check that the ESC you choose is compatible with your setup. 2S-3S, 2S-6S like the picture below...etc

SimonK is a firmware is preferred for multicopters.

Good stuff: Castle, Hobbywing, T-Motor

Make sure you calibrate your ESC's. Calibration ensures that all your motors engage in concert. See your ESC and flight controller manual for details. Here is 3DR's procedure: http://copter.ardupilot.com/wiki/initial-setup/esc-motor/

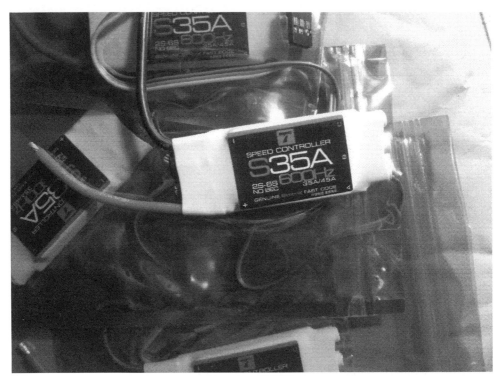

Tmotor ESC's

Propellers

Sizes - a 5030 prop is 5.0" long and has a 3.0" pitch.

Pitch - in one revolution a 5030 prop will move forward 3.0 inches.

Materials – Carbon Fiber, Beech Wood, Plastic. The stiffer the better.

Tip: Larger props are quieter.

Tip: motor/prop/battery matching is key.

Tip: Good stuff: HQ, APC, Aeronaut, Rctimer, Gemfan

Tip: There is no need to use a "pusher" prop on your airplane if your motor is rear mounted like a Bixler or flying wing. Pusher props are only for gas engines that are not reversible. Electric brushless motors are reversible by swapping two of the three power leads.

Tip: replace "prop-nuts" with locking nuts like the picture below, adds a lot more safety. These are 4mm that I got at the hardware store for about 40 cents.

Props break, make sure you have extras. Like most people I get miffed about broken props, but I have to remind myself that a well-designed prop is supposed to be the weak link and break, preventing damage to your motor and frame.

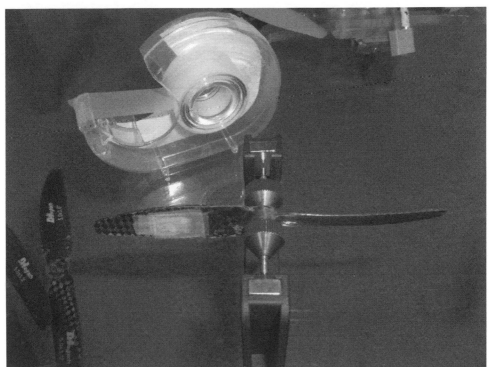

All new propellers require balancing. An unbalanced prop can make you video "jello" and cause wear on the motor bearings. For best results balance a prop with a purpose made magnetic stand. You can either add tape to the prop or take material off the heavy side with sand paper. This 5030 carbon prop took 5 layers of tape to balance, that's too much. I'll need to re-do this prop and sand down the other side. (http://amzn.to/1vIJ6Ed)

Chapter 8: Pre-Flight & Flying

Now that you have your FPV plane assembled, this chapter is for everything that you will need to check before flight (or the night before flight) with some techniques to get you started on your maiden flight.

Center of Gravity

Flying Wing and Delta Wings

The CG of a flying wing is of utmost importance to get correct on a flying wing because there is no tail to trim it out. So you have to do it before flight. Even a ¼ inch of battery positioning can make a huge difference. Flying wings don't have much lift in the front half so they will end up overly nose heavy if you use conventional formulas. Try the calculator below and follow your airframe manufacturer's directions on setting CG.

Flying wing CG calculator: http://fwcg.3dzone.dk/

RiteWing recommended CG for the ZII: "*CG is located on the bottom 9 3/8" to 9 1/2" from the nose back*". Slight nose heavy. Also can be measured as 3.5" from far rear of motor mount.

Conventional Aircraft

If you don't have manufacturer's directions then find the thickest part of the wing or find the spar, start there then measure back 1-1.5" and mark with a pencil. Then attempt to balance the plane on that location with your index fingers. Move battery or other payload around until it balances.

Setup Notes:
Aerobatic plane = neutral to slight tail heavy, dangerous!
Normal plane = slight nose heavy
Warplane = nose heavy

Multirotor

Yes CG is important on a quad too! Don't mount your camera gimbal or GoPro too far forward or the front motors will be working overtime resulting in a pitching sloppy quad.
...And don't forget about the propellers, they have a CG too!

100

Pre-Flight Systems Check

Checklists should always be evaluated for relevance and tailored to you.

Day **before** flight, check these items:

- Check Center of Gravity with all equipment installed, balance props.
- Batteries charged in all equipment. Radio, camera, main batteries, laptop etc.
- Preload Google Earth maps for offline use, 40 square miles?
- Plan the take-off location and primary and alternate landing sites.
- Note no-fly areas/airports.
- Memory cards and laptop are ready for recording. Plenty of gigabytes remaining?
- Check plane and hardware for damage from previous flights.
- Check weather forecast.
- Check and print out most current VFR chart at: vfrmap.com or skyvector.com
- Check for NOTAM at: pilotweb.nas.faa.gov/PilotWeb/

Day of, Pre-Flight checks:

- Check environmental frequencies with a Software Defined Radio (SDR) for interference on your video and radio frequencies.
- Unpack and set-up all equipment, while doing so check cables and connectors for damage.
- Ensure all connections are tight and secure. Batteries, antennas.
- Power up all systems. Note any magic smoke being vented, replace as necessary.
- Orient any directional antennas in direction of flight path.
- Check all functionality. Motors and control surfaces respond correctly. Video working? Prop balanced?
- Check GPS has a lock, 6-8 satellites is fair. 9+ is better.
- Calibrate plane and Gyro relationship to North azimuth.
- Weather is good?
- Turn on scanner, set to your local aviation-band frequencies. Ex: 118–136.975 MHz
- check flightradar24.com or app for semi-real time flight info, or use ADS-B/sdr receiver (just for fun, can be inaccurate info)
- Start cameras recording. Lenses clean?
- Start Timer.
- Checks complete, proceed with launch.

Basic flight rules

There will be a responsibility when flying because damage to persons or property could be a possibility, so don't fly where there are people or property! It is our duty to always fly in a safe and responsible manner. Considering that this is a relatively new hobby that is misunderstood by the media and legislators, you don't want to be the guy that gets made an example of. The media and the lawyers won't be friendly if your plane crashes into someone or something. There are people itching to ban and limit civilian use of UAV's, don't give them any fuel for their fire – be responsible and also follow your local laws.

Use a spotter to watch for other aircraft, the spotter should also be able to take control of your aircraft in case you have a medical emergency or other problems. It's always more enjoyable to fly with friends anyway.

It is easy to get disoriented when flying FPV, using GPS and Google Earth as well as your spotter is a good way to keep track of where your plane is *(or was)*.

Don't push your range on the first flight. Only go for long distance flights after you have the skill and have a proven and tested plane that has been on several short flights.

Get to know your plane and environment. Every time you fly it will be different.

Inclement weather can affect how your plane flies and the affect the range of your radio. Heavy clouds, rain and trees can block frequencies. But even in good weather your radio spectrum could get polluted by other people on your frequency, something as innocent as a long haul trucker passing by could wipe out your comms, some of these guys notoriously operate illegal linear amplifiers on their CB's which can spill over across and jam your frequency.

Environmental Restrictions

This chapter will highlight some of the restrictions that Mother Nature or Humans could use against your FPV RC airplane while in flight.

Inclement Weather - High winds, rain, ice, hail and fog will ruin your day. You can't do much about it either. Go play a flight simulator instead. You can use what's called a "Trainer Plug" or "Simulator Cable" which plugs your RC transmitter into your computer. Each controller has a different style of connector, contact your manufacture for the correct cable. Some use USB and some use what look like earphone connectors. These both allow you to use your actual RC controller as a game controller. With one of these cables, you can play Flight Simulator X, X-Plane or Flight Gear and get effective practice time in with your real RC controller. Or get a little $20 nano-quadcopter and fly around your living room! (http://amzn.to/YygzQO) (http://amzn.to/1nbavz8)

Distractions – People around you while you are flying.

Audible Noise - A good UAV shouldn't be noticed. A noisy plane draws attention and makes neighbors angry which compels them to use their shotguns on your plane. Things you can do to reduce noise; fly higher. And use larger propellers but at slower speeds, for instance the 15" props on the quad in the recipe at the back of the book is almost totally silent and only heard when it's about 20' overhead.

Hacking (purposely or accidentally):

> **RC take-over / co-opt the operations** – This used to happen all the time back in the day, whenever a group of guys would get together and fly RC airplanes, it never failed that someone else was on the same channel that you were using then you have to swap crystals in your radio. All new modern RC radios have Frequency Hopping Spread Spectrum (FHSS).
>
> RC radio protocols are relevantly simple and could be emulated and co-opted by an attacker that has a transmitter that is more powerful than yours and the ability to clone your PPM/PWM signal, essentially pushing you out of the way. The potential to "steal" or purposely crash your plane is remote because this tech is very sophisticated.
>
> **Video channel surfing** – If you are broadcasting un-encrypted video, it is also available to everyone else that has the equipment to view it. A security risk? FlyTron has a tiny analog FPV video scrambler, like the old cable TV scrambled channels. While not encrypted, it is obfuscated. $120

Real encryption is available with Dji's Lightbridge wifi HD video link. But, if you lose connection it takes some time to come back. During that black out time you can only hope you are in the air still.

GPS Jamming - You can still fly FPV without GPS, but you will lose return-to-home and autopilot functions. Hone up on your manual directional navigation. Many new GPS modules feature enhanced jamming detection and negation, such as the uBlox 8M.

Laser Weapon System – (LaWS) the US Navy has a system that can burn UAV's from the sky! Cool video link! (http://youtu.be/OmoldX1wKYQ)

EM Environmental Pollution - Use an SDR with uni-directional antenna to DF locate frequencies, then adjust your system accordingly. Some ways of negating EM noise: Component placement, Antenna placement, Shielding, Twisted pairs, Ferrite cores, Notch filters.

Directed Energy/EMP (Nature or human derived):

> **EMP** – Electromagnetic pulse. A term that prompt's tinfoil hatters to double wrap. It's very unlikely to happen... but it has; in the year 1859, a solar flare occurred that was a super EMP that melted telegraph lines. Nuclear weapon detonations can also create large area of effect EMP's, but that's the least of your problems at that point.
>
>> **Boeing's CHAMP weapon** – A new directed EMP technology that is mounted to a UAS that advertises the ability to shut down enemy drones.
>
> **Directed microwave RF energy weapon** - There are rumors of a few guys on the internet that have built (illegally) such weapons out of microwave ovens and large capacitors. They purportedly were able to disable a car that was annoyingly bumping his stereo a bit too loud. I'm not sure of the truth of this or the effective range. The internet is full of poor information on this, beware. Also highly dangerous to humans.

Birdshot - A skilled shot gunner can hit a moving target at 300'. Recently in the news; an environmentalist group found this out the hard way when they sent their FPV Quadrotor to go spy on a bird gun club packed full of guys with shotguns. The club ended up using the quadrotor for target practice.

Failure Management

If you lose more than one, you may be done.

These are just suggestions, you will need to figure out what to do with your system configuration.

Loss of GPS - Minor casualty. Don't click on RTH or Auto or Loiter, it will crash. Manually fly home via FPV in stabilize (no GPS) mode.

Drivetrain/prop failure - Moderate casualty. Glide to an uninhabited field so that your FPV doesn't cause damage or get stolen by the locals. Major failure for a quadcopter, however an octocoper can sustain one or two motor losses and continue flight.

Loss of FPV video link - Moderate to Major casualty. If the loss was due to it flying behind a RF barrier -Try a full throttle climb, might bring your plane into line of sight link range. In not, then enable RTH and pray.

Loss of RC - Major casualty. Plane is gone unless RTH is allowed to auto-enable upon throttle loss PWM failsafe. Plot direction plane was going. Hopefully you have a GPS transponder, GTU-10.

Servo failure/stuck aileron - Major casualty. Try and limp it home or find a safe spot to crash land. Things happen fast so be ready.

Energy failure - Major casualty. Plot direction and go find it.

Chapter 9: Mods and Fun Stuff!

Backup GPS tracker - These are usually used for tracking pet dogs, vehicles or cheating spouses. They send location over cellular networks and some use GMRS/FRS. Here is one that lightweight (1.8oz) that uses cellular to email its position to you, includes 1yr of cell service and then its $50 each year after: Garmin GTU 10 GPS Tracking Unit. If you want something smaller; there is the Tagg GPS pet tracker that is smaller and lighter (1.1oz), but has a monthly fee of $7.99 after 3 months free. If you fly where there is no cellular service, then you might look into a "direction finding" system. The kind that are used for tracking wild animals. (http://amzn.to/10lFfT2) (http://amzn.to/1bKgjqG)

Note: The Garmin and the Tagg both use AT&T service, while other cheaper trackers 'China specials' use Euro GSM or US T-Mobile. Choose the one that has the best coverage in your flying area.

Garmin GTU-10

Folding Wing / 2 Piece wing - Makes it easy to store and transport in the trunk of a car. Electrical connections, wires and antenna placement will have to be carefully considered. Also it may sacrifice rigidity for portability.

Diversity – Automatic video switching between antennas. Try using two different frequencies and/or antenna styles, on a long range flight you get a sense of how much further you can go when the higher frequency link drops out and then

106

transitions to your lower frequency antenna. Or another way to use it is to use two or more directional antennas aimed so that the beam edges slightly overlap and switch between them to give you more degrees of coverage, that's essentially better than an "Antenna Tracker" because it doesn't rely on GPS and Telemetry but simply quality of signal. EzUHF has onboard diversity as an option, really cool.

Antenna tracking – Should be used for dish antennas or really directional antennas. I feel it is over used and diversity is better and more reliable, but if must have the coolness of a robotic tracking antenna then read on: Tracking is a technologically sophisticated way to aim your antenna at your plane (using Diversity is generally preferred over Tracking). Here's how it works; onboard GPS sends a signal to the ground station which also has a GPS, the ground station box or laptop that does some math which then outputs to some servos that aim the ground station unidirectional antenna at the omni-directional UAV in the air and adjusts on the fly as the UAV flies around.

There are commercial versions of these but they have a spotty reliability record. Don't rely on antenna tracking to always work. Just be ready to manually take over aiming if it fails. GPS loss or telemetry loss will negatively affect your tracking, *"Crash!"*.

If you had a large enough airplane (think huge) you could do the whole thing in reverse too. Put a tracker'ed unidirectional antenna on the drone with a radome, it would dramatically increase range and information security (unidirectional antennas have higher dB and a narrow beam width compared to an omni-antenna).

Waterproofing - Humiseal 1b31, Electronics water proofing spray. If in a hurry I also wrap all bare electronics in Kapton tape (polyamide tape), makes it semi-water resistant. (http://amzn.to/1s37TDT)

Launching systems – Slingshot or catapult. Safer than hand launching. Slingshots are easy and portable, catapults are large and may even need a trailer. If you get tired of hand launching consider getting a slingshot. Just make sure the bungee anchor is well set or else you could get a steel anchor to the head.

Gimbaled camera mount – These are usually more complex than what we need in FPV, but it is cool to steer your camera around or use in conjunction with a gyro for stabilize the video. Use a brushless gimbal for ultra-smooth movement, or use servos if it needs to be cheaper. Also look into "Geo pointing" for the lat-long of what your camera is aimed at.

Zoom Lens – Most consumer video cameras come with high quality 10-20-40x optical zoom lenses. Why not strip one down to the lens and camera, you probably already have one in your basement that you only used once for that wedding or birthday party. Pair this with a gyro'ed gimbal mount for stability and aiming. This would be the first step towards full-on UAV high altitude surveillance capability. Make sure it has a video out connector and disable auto-off. Size and weight might be an issue so it should be stripped to its basics if possible. **Bonus**: Some of these cameras have an IR "night" mode, which would make a nice "poor" man's night vision. Disable or cover any IR LED illuminators as they will illuminate close clouds and fog which is a hindrance.

Laser beam communication - This is more pie-in-the-sky than practical, but for fun: A mounted invisible IR laser that could be used as a high speed semi-secure communications link. Mounting would have to have gimbaled with auto tracking. The most difficult part might be the aiming and the receivers, both would have to be very sensitive.

Panoramic camera – There are a few ways to do this: Mount several cameras and later stitch the pictures together in a photo editor, or buy a true panoramic camera. **Experiment**: I haven't tried this, but maybe you will. Instead of buying a dedicated panorama camera, mount a GoPro underneath the plane and hack one of these inexpensive cellphone Kogeto Dot lenses to get some 360 degree panoramic pics or video! Kogeto is reportedly working on a lens specifically for the GoPro. A gimbaled gyro mount would be recommended for crisper pictures. (http://amzn.to/176ID3c)

Night Vision and Thermal – Real night vision costs thousands of dollars, and would require at least a Gen 3 monocular + camera interface. But for real cheap you can get close to good NV with a Sony grayscale CCD camera that goes down to .0003 lux or lower. Thermal is better than NV, but it usually has lower frame rates of 15fps, so a super stable airframe and gimbal would be required. If you have cash to burn these are super small modules: Tamarisk 320 or the FLIR Tau 2 and FLIR Quark. Costs are 3-5k! ouch! (http://bit.ly/1sdx9Gb) (http://bit.ly/1nl7WdS) (http://bit.ly/1xQbMN8)

Droppable payloads - Burrito/Taco delivery. Do an internet video search for the *taco drone* and *burrito bomber*, both are hilarious. But could also be used for real things: dropping medical supplies or emergency water to hikers lost in the desert. Hobbyking's "Candy Dropper" would be an easy way to get going.

Lighting - Either running lights or decorative, LED or Electroluminescent wire. There are kits made specifically for RC aircraft. Maybe rig up some animated LED array to look like a UFO, freak out the local trailer park.

Solar power – Yes you can add solar to charge while flying, however the solar cells are really heavy and are not efficient enough to charge enough to maintain flight unless you have a super large wingspan, probably football stadium sized wingspan.

If you can pull this off and accomplish 24-7 flight, you win the Engineer of the decade award from me.

3D FPV flight cam for full immersion - Use two cameras and two video transmitters. Space the onboard cameras to your Pupillary Distance (from pupil to pupil), if it isn't set correctly then headaches and dizziness may occur. Usually around 2.5" wide, but even a small error of 1/8" will cause headaches. Use FPV glasses that have two separate LCD's with individual inputs. Might also require two video transmitters on different frequencies or a reduced resolution on one freq. Oculus Rift...

3D HD secondary camera – The Panasonic DMC-3D1 or the Fujifilm Finepix Real 3D are nice small point and shoot style cameras that do 3D HD video and stills. Or you can add a 3D lens to a micro four thirds camera or camcorder. The inexpensive Aiptek 3D-HD might be an option too if you can fit its odd tall shape into your fuselage. (http://amzn.to/18JmUlj) (http://amzn.to/10JB2WV) (http://amzn.to/1940A6h)

Twin engines - More speed and power! But more weight. Wire separately to gain some redundancy. WWII style P38 Lightning FPV, sign me up.

Fuel Cell – (see in Chapter 5):

Custom parts 3D printing - Get creative and make your own parts. Maybe a gimbal camera mount for your 2nd camera. Thingiverse.com has gimbal CAD files that are ready to print in plastic.

AI - Program some Artificial Intelligence then share with the community. Let's bring *Skynet* online! I, for one, welcome our new robot overlords.

Parachute - Parachutes can be a gentle way to crash land your plane, errr *safely* if they work. They are commercially available for RC aircraft. They are not without problems though, sometimes they don't deploy and sometimes they deploy when you don't want them to, doh.

Head tracker - Head tracking is a virtual reality technology. It's a tiny accelerometer and transmitter that mounts to your video goggle head gear, so when you move your head left or right or up and down it sends a signal to your plane camera gimbal servos. So it feels like you are in the plane and can look around freely. Sounds cool, but adds weight and is somewhat of a novelty and impractical. Also, if you gimbal a camera it shouldn't be your sole camera onboard because if you lose your gimbal control link then your camera could be facing the opposite direction than you need, then you crash.

Graphics - I like to see RC planes that are painted blaze orange or white for safety and recovery, and it's easier to fly what you can see from the ground. Dark colors absorb heat from the sun more, so choose lighter colors. Camo isn't recommended on FPV's because it makes your plane hard to find if it goes down and is a potential danger to other aircraft.

Repeater link:

> **Satellite repeater** - Your own Satellite may sound like wishful thinking, but it can be done. There are actually amateur satellites up there in orbit right now. You can build a simple satellite and rent a ride into outer space from the Jet Propulsion Laboratory or NASA, cost is/was about $10,000 per pound. It *might* be possible to do something similar with a cheaper *satellite phone* with a data plan or modem. A potential problem is aiming the onboard antenna and data throughput and bandwidth rates.

> **Cellular network link** - I can't fully recommended doing this because when your plane is hauling ass across the sky it will keep hopping cell towers, resulting it breaks in your signal. But if you have good cell coverage, it might work. It could be neat to use it as an augmentation to your existing FPV radio gear. *Dronecell* is an inexpensive purpose built system for 2-way GSM telemetry and control.

> **Balloon repeater link** - Use a tethered helium or hot air balloon at 100' (?) or a high altitude 100,000"+ near space balloon….cheap faux satellite! which could carry a WiFi repeater up and over obstructions. This would be great if you are in a wooded area where you don't have good radio line-of-sight to the drone, could significantly increase your range. Be sure to plan your balloon flight with the FAA.

> **Tower repeater** – Alternatively, and probably the easiest of all you could build a tower, imagine it like a cell tower or HAM repeater radio tower. Would be ideal if you had a farm and flew regularly to keep tabs on your land and livestock.

Chapter 10: Custom Drone Recipes

There are a million different combinations that could be used to make a drone. But what works? Here are 8 ways that work, three planes and five quads. (the 650mm recipe is actually three recipes)

For the purpose of lightness; the diagrams feature shortened wire lengths with common rail architecture that requires soldering and a bunch of time. You don't have to take this option.

No warranty or guarantee!

FPV 30" Mini Z flying wing, $592

This setup is a small and packable FPV and for $592 total will get you a few miles with 35 minutes of run time. MUCH easier to hand launch than the 55". Con is low payload capacity, but its nimble and fast.

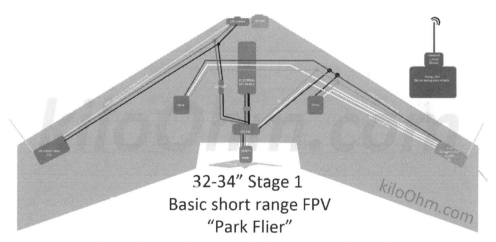

32-34" Stage 1
Basic short range FPV
"Park Flier"

Large image link (http://kiloohm.com/?p=483)

Airframe – RiteWing Mini Z 30-34"", $59 readymaderc.com, ritewing.com
    Alt: Homebuilt from hardware store foam or Depron, Popwing 35", Teksumo 900mm
VTX – 1.3 GHz set to 1280MHz 1000mW, $89 SierraRC.com, securitycamera2000.com
    Alt: 5.8GHz Fatshark
VRX – 1.3 GHz Upgraded Saw set to 1280MHz, $44 SierraRC.com, securitycamera2000.com

Alt: 5.8GHz Fatshark

LRS – 2.4GHz radio if not going long range, FRsky D4R-II $24 with FrSky XJT $38 (9xr needs the XJT, because it doesn't come with a 2.4 module)

Alt: LRS 433MHz 1W, Hawkeye DTF UHF (OpenLRS), dipole, $120 for RCtx and RCrx, multirotorsuperstore.com

Video display – Haire 7" LCD (no-blue screen), eBay $40

Flight Camera – RMRC-mini-V2, $40 readymaderc.com

Alt: Fatshark cam

12v for cam - none because we are using 2S batts, this tiny cam is 5-6v

Video Power Filter - L-C Type $12 dpcav.com (or use toroid)

Propeller – APC 8x5 Electric Prop, $4 atlantahobby.com

Covering - $20 readymaderc.com, ritewing.com

RC transmitter – Turnigy 9XR mode 2, $50 hobbyking.com

Alt: FrSky Taranis

Transmitter battery – Turnigy Safety Protected 11.1v 3s 2200mAh 1.5C, $15 hobbyking.com

Motor – Tmotor MT2206-13 2000kv $45

Alt: Sunnysky X2208 2350kv, Sunnysky, x2204-21, rctimer A2208-12

Speed Control – Blue 12A ESC, $9 hobbyking.com

Alt: Tmotor 12A or Hobbyking Plush 12A or 18A, Castle 18A

Battery – 2S Nano-Tech 3300mAh 25-50C, $15ea total hobbyking.com

Connectors – XT60, set of 5 pairs, $3 hobbyking.com

Charger – B6AC, $35

Ground Station 12v Battery – 3S 5000mAh, B Grade $8, hobbyking.com

Servos – two: Hitec HS-65HB, 9-11 gram $40 total hobbyking.com

Alt: HS-5065HB, HTX900

Servo parts:

CNC Alloy Control Horns, $7

Glass Fiber Rod, for pushrods 2.0x750mm, $1

Alloy Clevis for Non Threaded 2mm Control Rod, $3 x2

Extra Junk you may or may not need

Rare-Earth Button Magnets for lids (10), $1, x2 hobbyking.com

Heat Shrink Tube – 4mm, 1 foot, $0.21 hobbyking.com

Battery straps (4), $4 hobbyking.com

Polyester peel and stick Velcro (1mtr), $2

Kapton Tape, $5 eBay

Gorilla Glue, $5 home improvement store

Extreme Tape, $5 home improvement store

Alt: Storage Tape

Hot glue, $1 "hair extension" high temp hot glue polyamide.

Electrical Tape to wrap servos, $1 home improvement store

Blue Loctite, $2 auto parts store

Additional Options for experienced flyers

HD Camera - GoPro 3 – Amazon

HD Camera microSD card – PNY 32GB

VTX Antenna – 1.3 GHz Skew Planar RHCP, SierraRC.com

VRX Antenna – 1.3 GHz Helical or Crosshair RHCP, SierraRC.com

RCRX Antenna – dipole, diy

RCTX Antenna – $50 arrowantennas.com 440-7ii

Notch Filter for LRS - IBCrazy 433/1.3, $20 readymaderc.com

VRX Diversity Controller + additional antenna -

Autopilot – Ardupilot APM 2.5 or 2.6

Telemetry – 3DR 900MHz/100mw, or use EzOSD or TinyTelemetry audio downlink.

　　　Alt: 2 Watt amplifier/booster, shireeninc.com/900-mhz-2-watts-oem-module-amplifier/

　　　Alt: RFD900 1 watt radios

OSD – EzOSD $175

　　　Alt: minimOSD $15

Current Sensor – free with EzOSD, 3drobotics power module

GPS – free with EzOSD

GPS Cellular Locator - Garmin GTU 10 Tracking Unit.

Battery upgrade –

Ground Control Case –

Video Splitter -

PC USB analog video input or DVR -

USB RTL SDR – **R820T**, $20.

Laptop to run mapping.

Tripod – $30

FPV Headset – Fatshark?

Extra Batteries –

Extra Chargers -

Extra Props - these do eat props on landing

FPV 54" Zephyr Flying Wing, $691

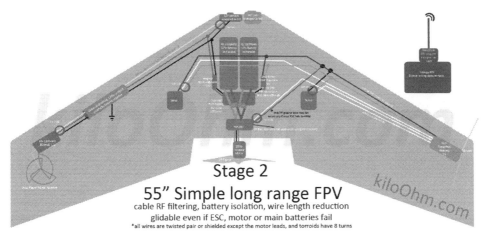

Stage 2
55" Simple long range FPV
cable RF filtering, battery isolation, wire length reduction
glidable even if ESC, motor or main batteries fail
*all wires are twisted pair or shielded except the motor leads, and torroids have 8 turns

Large image link (http://kiloohm.com/?p=483)

This setup is the middle ground FPV and for $691 total will get you to about 5-7 miles safely, by updating the ground antennas to directional will grant you 20-40 miles. Just add autopilot/osd for a full on UAV (stage 3 on the next page). Can carry a couple pounds.

Airframe – RiteWing Zephyr II 54", $129 readymaderc.com, ritewing.com
        Alt: Homebuilt from hardware store foam or Depron
VTX – 1.3 GHz set to 1280MHz 1000mW, $89 SierraRC.com, securitycamera2000.com
VRX – 1.3 GHz Upgraded Saw set to 1280MHz, $44 SierraRC.com, securitycamera2000.com
LRS – 433MHz 1W, Hawkeye DTF UHF (OpenLRS), dipole, $120 for RCtx and RCrx, multirotorsuperstore.com
Video display – Haire 7" LCD (no-blue screen), eBay $40
Flight Camera – PZ0420 2.8mm, 12v, $45 securitycamera2000.com
        Alt: CMQ1993x
12v for cam and VTx- Adjustable Voltage Regulator, 1-35V SEPIC Type $12 SierraRC.com, dpcav.com
        Alt: just tap off the battery balance plug
Video Power Filter - L-C Type $12 dpcav.com (or use toroids)
Propeller – APC 10x5 Electric Prop, $4 atlantahobby.com
        Alt: Folding Prop Set 10x6 Aeronaut CAM, team-blacksheep.com
Covering - $20 readymaderc.com, ritewing.com
RC transmitter – Turnigy 9XR mode 2, $50 hobbyking.com
        Alt: FrSky Taranis

Transmitter battery – Turnigy Safety Protected 11.1v 3s 2200mAh 1.5C, $15 hobbyking.com

Motor – NTM Prop Drive Series 35-42A 1250Kv 600W, $20 hobbyking.com

>>Requires: NTM Prop Drive 35 Series Accessory Pack, $2
>>Alt: Turnigy Aerodrive SK3 - 3542-1250kv

Speed Control – Turnigy AE-80A Brushless ESC, $35 hobbyking.com

Motor Batteries – two 4S NanoTech 3300mAh 25-50C in parallel, $70 total hobbyking.com

Connectors – XT60, set of 5 pairs, $3 hobbyking.com

Charger – B6AC, $35

Ground Station 12v Battery – 3S 5000mAh, B Grade $8, hobbyking.com

Servos – two: Hitec 645mg $60 total hobbyking.com

Servo parts:

>CNC Alloy Control Horns, $7
>Glass Fiber Rod, for pushrods 2.0x750mm, $1
>Alloy Clevis for Non Threaded 2mm Control Rod, $3 x2

Extra Junk you may or may not need

>Rare-Earth Button Magnets for lids (10), $1, x2 hobbyking.com
>Heat Shrink Tube – 4mm, 1 foot, $0.21 hobbyking.com
>Battery straps (4), $4 hobbyking.com
>Polyester peel and stick Velcro (1mtr), $2
>Kapton Tape, $5 eBay
>Gorilla Glue, $5 home improvement store
>Extreme Tape, $5 home improvement store
>>Alt: Storage Tape
>Hot glue, $1 "hair extension" high temp hot glue polyamide.
>Electrical Tape to wrap servos, $1 home improvement store
>Blue Loctite, $2 auto parts store

<u>Additional Options for experienced flyers</u>

HD Camera - GoPro 3 – Amazon

HD Camera microSD card – PNY 32GB

VTX Antenna – 1.3 GHz Skew Planar RHCP, SierraRC.com

VRX Antenna – 1.3 GHz Helical or Crosshair RHCP, SierraRC.com

RCRX Antenna – dipole, diy

RCTX Antenna – $50 arrowantennas.com 440-7ii

Notch Filter for LRS - IBCrazy 433/1.3, $20 readymaderc.com

VRX Diversity Controller + additional antenna -

Autopilot – Ardupilot APM 2.5 or 2.6

Telemetry – 3DR 900MHz/100mw, or use EzOSD or TinyTelemetry audio downlink.

Alt: 2 Watt amplifier/booster, shireeninc.com/900-mhz-2-watts-oem-module-amplifier/

Alt: RFD900 1 watt radios

OSD – EzOSD $175

Alt: minimOSD $15

Current Sensor – free with EzOSD, 3drobotics power module

GPS – free with EzOSD

GPS Cellular Locator - Garmin GTU 10 Tracking Unit.

Battery upgrade –

Ground Control Case –

Video Splitter -

PC USB analog video input or DVR -

USB RTL SDR – **R820T**, $20.

Laptop to run mapping.

Tripod – $30

FPV Headset – Fatshark?

Extra Batteries –

Extra Chargers –

Extra Props - these do eat props on landing

FPV/UAS 54" Zephyr Flying Wing, $876

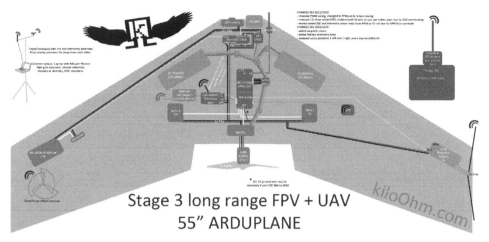

Stage 3 long range FPV + UAV
55" ARDUPLANE

large image link (http://kiloohm.com/?p=483)

This setup is a full on FPV *and* UAS. Batteries have to relocated to the sides to make room for the autopilot.

Autopilot – Ardupilot APM 2.5 or 2.6, 900MHz/100mW telemetry, Ublox, $120
  rctimer.com, 3drobotics.com
Airspeed Sensor Pitot – $25 3drobotics.com
Current sensor "APM Power Module" - $25, 3drobotics.com
OSD – minimOSD $15, hobbyking.com, 3drobotics.com, etc. Beware, don't wire up 12v. Only 5v on both sides.
All items below are the same as the $691 FPV 54" build
Airframe – RiteWing Zephyr II 54", $129 readymaderc.com, ritewing.com
  Alt: Homebuilt from hardware store foam or Depron
VTX – 1.3 GHz set to 1280MHz 1000mW, $89 SierraRC.com, securitycamera2000.com
VRX – 1.3 GHz Upgraded Saw set to 1280MHz, $44 SierraRC.com, securitycamera2000.com
LRS – 433MHz 1W, Hawkeye DTF UHF (OpenLRS), dipole, $120 for RCtx and RCrx, multirotorsuperstore.com
Video display – Haire 7" LCD (no-blue screen), eBay $40
Flight Camera – PZ0420 2.8mm, 12v, $45 securitycamera2000.com
  Alt: CMQ1993x
12v for cam and VTx- Adjustable Voltage Regulator, 1-35V SEPIC Type $12 SierraRC.com, dpcav.com

Alt: just tap off the battery balance plug

Video Power Filter - L-C Type $12 dpcav.com (or use toroids)

Propeller – APC 10x5 Electric Prop, $4 atlantahobby.com

Alt: Folding Prop Set 10x6 Aeronaut CAM, team-blacksheep.com

Covering - $20 readymaderc.com, ritewing.com

RC transmitter – Turnigy 9XR mode 2, $50 hobbyking.com

Alt: FrSky Taranis

Transmitter battery – Turnigy Safety Protected 11.1v 3s 2200mAh 1.5C, $15 hobbyking.com

Motor – NTM Prop Drive Series 35-42A 1250Kv 600W, $20 hobbyking.com

Requires: NTM Prop Drive 35 Series Accessory Pack, $2

Alt: Turnigy Aerodrive SK3 - 3542-1250kv

Speed Control – Turnigy AE-80A Brushless ESC, $35 hobbyking.com

Motor Batteries – two 4S NanoTech 3300mAh 25-50C in parallel, $70 total hobbyking.com

Connectors – XT60, set of 5 pairs, $3 hobbyking.com

Charger – B6AC, $35

Ground Station 12v Battery – 3S 5000mAh, B Grade $8, hobbyking.com

Servos – two: Hitec 645mg $60 total hobbyking.com

Servo parts:

CNC Alloy Control Horns, $7

Glass Fiber Rod, for pushrods 2.0x750mm, $1

Alloy Clevis for Non Threaded 2mm Control Rod, $3 x2

Extra Junk you may or may not need

Rare-Earth Button Magnets for lids (10), $1, x2 hobbyking.com

Heat Shrink Tube – 4mm, 1 foot, $0.21 hobbyking.com

Battery straps (4), $4 hobbyking.com

Polyester peel and stick Velcro (1mtr), $2

Kapton Tape, $5 eBay

Gorilla Glue, $5 home improvement store

Extreme Tape, $5 home improvement store

Alt: Storage Tape

Hot glue, $1 "hair extension" high temp hot glue polyamide.

Electrical Tape to wrap servos, $1 home improvement store

Blue Loctite, $2 auto parts store

<u>Additional Options for experienced flyers</u>

HD Camera - GoPro 3 – Amazon

HD Camera microSD card – PNY 32GB

VTX Antenna – 1.3 GHz Skew Planar RHCP, SierraRC.com

VRX Antenna – 1.3 GHz Helical or Crosshair RHCP, SierraRC.com

RCRX Antenna – dipole, diy

RCTX Antenna – $50 arrowantennas.com 440-7ii

Notch Filter for LRS - IBCrazy 433/1.3, $20 readymaderc.com

VRX Diversity Controller + additional antenna -

Telemetry booster - 2 Watt amplifier/booster, shireeninc.com/900-mhz-2-watts-oem-module-amplifier/

   Alt: RFD900 1 watt radios

Alt: OpenLRS 2 way 1W

GPS Cellular Locator - Garmin GTU 10 Tracking Unit.

Battery upgrade –

Ground Control Case –

Video Splitter -

PC USB analog video input or DVR -

USB RTL SDR – **R820T**, $20.

Laptop to run mapping.

Tripod – $30

FPV Headset – Fatshark?

Extra Batteries –

Extra Chargers -

Extra Props – these do eat props on landing

FPV 250mm racing Mini-H-Quad 3S 5" props, $500-750

This setup is a small "Blackout" style 250mm racing FPV quad, tons of fun! 5-10 minutes flight. Could be made cheaper without the LRS and 1.3 system. Could use 2.4GHz and 5.8GHz just fine for a third of the cost, but I like to fly in forests so I use LRS. I used a homemade frame made of Plexi-glass/Lexan. You can get the template here: (http://kiloohm.com/?p=461)

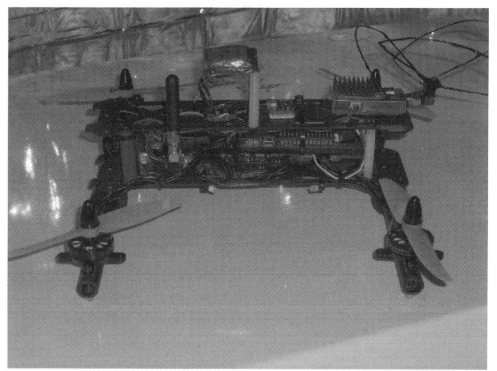

An APM based 280mm mini-H that I made and broke in half. It was not rigid enough to take a crash.

With 5030 props; 1880g of total thrust, keep the AUW below half that. 2840g of thrust with 6045 props. My plexiglass version is 500g AUW, a bit portly but strong. A carbon fiber version could be made under 350g AUW.

Autopilot – Acro Naze32 $20 multirotorsuperstore.com, I did install an APM in mine and it flew fine but the Naze was more locked in. The Naze32 is solid on this little frame. Check my website for APM PID settings.
http://kiloohm.com/pid-tuning-an-arducopter-apm-2-6-mini-h-quad/
Airspeed Sensor Pitot –
Current sensor -
OSD –
Airframe – Blackout 250mm $150 or Lumineer QAV250 $120 or ebay clone $40 or free!
VTX – 1.3 GHz set to 1280MHz 1000mW, $89 SierraRC.com, securitycamera2000.com
VRX – 1.3 GHz Upgraded Saw set to 1280MHz, $44 SierraRC.com, securitycamera2000.com
LRS – 433MHz 1W, Hawkeye DTF UHF (OpenLRS), dipole, $120 for RCtx and RCrx, multirotorsuperstore.com
Video display – Haire 7" LCD (no-blue screen), eBay $40

Flight Camera – PZ0420 2.8mm, 12v, $45 securitycamera2000.com
  Alt: CMQ1993x
12v for cam and VTx- use 3S batt
Video Power Filter - L-C Type $12 dpcav.com (or use toroids)
Propeller – 5030 Carbon, ebay or hobbyking, $10 for 4
RC transmitter – Turnigy 9XR mode 2, $50 hobbyking.com
  Alt: FrSky Taranis
Transmitter battery – Turnigy Safety Protected 11.1v 3s 2200mAh 1.5C, $15 hobbyking.com
Motor – Sunnysky 2300kv $20ea, alt Cobra 2300kv, tmotor 2300kv
Speed Control – Turnigy qbrain 20A with heat sinks removed $20, or any 20A esc's
Motor Batteries – : 3S Zippy Compact 1800mAh 25-30C, $15 hobbyking.com
Connectors – XT60, set of 5 pairs, $3 hobbyking.com
Charger – B6AC, $35
Ground Station 12v Battery – 3S 5000mAh, B Grade $8, hobbyking.com
Extra Junk you may or may not need
  Heat Shrink Tube – 4mm, 1 foot, $0.21 hobbyking.com
  Battery straps (4), $4 hobbyking.com
  Kapton Tape, $5 eBay
  Hot glue, $1 "hair extension" high temp hot glue polyamide.
  Blue Loctite, $2 auto parts store
Additional Options for experienced flyers
HD Camera - GoPro 3 – Amazon
HD Camera microSD card – PNY 32GB
VTX Antenna – 1.3 GHz Skew Planar RHCP, SierraRC.com
VRX Antenna – 1.3 GHz Helical or Crosshair RHCP, SierraRC.com
RCRX Antenna – dipole, diy
RCTX Antenna – $50 arrowantennas.com 440-7ii
Notch Filter for LRS - IBCrazy 433/1.3, $20 readymaderc.com
VRX Diversity Controller + additional antenna -
Telemetry booster -
GPS Cellular Locator - Garmin GTU 10 Tracking Unit.
Battery upgrade -
Ground Control Case –
Video Splitter -
PC USB analog video input or DVR -
USB RTL SDR – R820T, $20.
Laptop to run mapping.
Tripod –
FPV Headset – Fatshark?

Extra Batteries –
Extra Chargers -
Extra Props – You will need lots.

FPV/UAS 400mm Entry level APM quad, 3S 8" props, $645

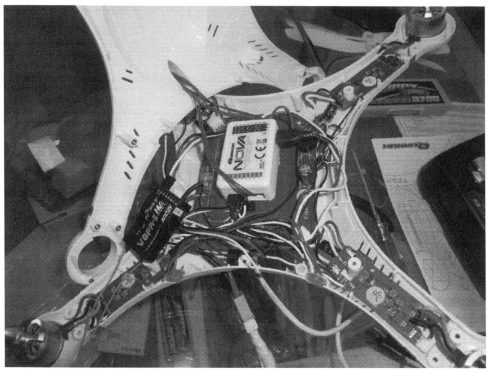

You shouldn't have to ever open this this thing up unless you need to replace an ESC. (note: the FrSKY was an upgrade)

This setup is easy to fly, entry sized almost ready to fly FPV/UAS quad with 15 minutes of real flight time and has full Arducopter Mission Planning and GPS. This recipe has a "all-in-one" FPV system so you just slap it on the quad. Could be cheaper and better if you rolled your own FPV system.

Autopilot – *mini*-APM 2.6 included (does not have a standard sized APM aux pin header, but that shouldn't matter for most)

Airspeed Sensor Pitot –

Current sensor "APM Power Module"-

OSD –

Airframe – Hobbyking Quanum Nova, with 2.4g radio. Just add battery and FPV. $319

　　　　Alt: Dji Phantom

VTX – FatShark PredatorV2 RTF FPV Headset System w/Camera and 5.8G TX $279

VRX – included

124

LRS –
Video display –
Flight Camera –
Gps mount -
12v for cam and VTx-
Video Power Filter -
Propeller – 8" included
RC transmitter –
Transmitter battery –
Motor –
Speed Control –
Motor Batteries – Zippy 3S 2800mah battery 111mm x 33mm x 25mm with XT60 connector $12
Connectors –
Charger – B6AC, $35
Ground Station 12v Battery –
Extra Junk you may or may not need – mini-USB cable, 3DR telemetry radios, gimbal.

Additional Options for experienced flyers

HD Camera - GoPro 3 – Amazon
HD Camera microSD card – PNY 32GB
VTX Antenna – circular 5.8GHz
VRX Antenna – circular 5.8GHz
RCRX Antenna –
RCTX Antenna –
Notch Filter for LRS -
VRX Diversity Controller + additional antenna -
Telemetry booster -
GPS Cellular Locator –
Battery upgrade –
Ground Control Case –
Video Splitter -
PC USB analog video input or DVR -
USB RTL SDR – R820T
Laptop to run mapping.
Tripod –
FPV Headset – included in fpv kit
Extra Batteries – !!!
Extra Chargers -
Extra Props - !!!

Note: this frame is also called a Cheersun CX-20. It has a two variants, one with a APM and one with a dumb flight controller (ZERO model).

FPV/UAS 650mm Big-Quad/Octocquad 4S, 15" props $999, or 1000mm Octocopter option

Payload is a loud speaker for a project I was working on.

This setup is a large FPV/UAS quad with 20 minutes of real flight time, hover time could be 35min. You could make it cheaper without the LRS and 1.3 system. Could use 2.4GHz and 5.4GHz just fine for a third of the cost, but I like to fly in forests so I use LRS.

I have over a hundred hours of flight time on this. If I built this again, I would not have put the ESC's underneath, I would have put them near the motors for less EM noise near the compass. I initially used an APM with internal magnetometer (compass) and had to transition to an external unit mounted near the GPS. Before it would freak out in auto mode and start doing circles at full speed.

BONUS: This recipe can also be an OCTOQUAD!!! Just double the motors, and ESC's and get 4 more motor plates from Tarot, use two batteries in parallel instead of one. **Double bonus:** and it can also be an **OCTOCOPTER** if you use an octo frame like the Tarot 1000!!!

All up weight without batteries or cameras: 1191g (2.6lb), which leaves about 700g for batts and cameras. (Which is quite lightweight for a 650 class quad with 15″ props.)

With 1555 props in 4S quadcopter config; 3800g of total thrust, keep the AUW below half that. More thrust can be added with larger props/pitch, more voltage or more motors. An octo with 6S and 17" props estimated to be around 12,000-16,000g thrust. 18-25kg thrust if motors are upgraded.

Autopilot – Ardupilot APM 2.5 or 2.6, 900MHz/100mW telemetry, Ublox, $120 rctimer.com, 3drobotics.com

Airspeed Sensor Pitot –

Current sensor "APM Power Module" - $25, 3drobotics.com

OSD – minimOSD $15, hobbyking.com, 3drobotics.com, etc. Caution: don't wire up 12v. Only 5v on both sides.

Airframe – Tarot Ironman 650mm $115 amazon

VTX – 1.3 GHz set to 1280MHz 1000mW, $89 SierraRC.com, securitycamera2000.com

VRX – 1.3 GHz Upgraded Saw set to 1280MHz, $44 SierraRC.com, securitycamera2000.com

LRS – 433MHz 1W, Hawkeye DTF UHF (OpenLRS), dipole, $120 for RCtx and RCrx, multirotorsuperstore.com

Video display – Haire 7" LCD (no-blue screen), eBay $40

Flight Camera – PZ0420 2.8mm, 12v, $45 securitycamera2000.com
     Alt: CMQ1993x

Gps mount - Foldable GPS mount $12

12v for cam and VTx- Adjustable Voltage Regulator, 1-35V SEPIC Type $12 SierraRC.com, dpcav.com

Video Power Filter - L-C Type $12 dpcav.com (or use toroids)

Propeller – 15x5.5" Carbon, Rctimer, or 16" $25 for four

RC transmitter – Turnigy 9XR mode 2, $50 hobbyking.com
     Alt: FrSky Taranis

Transmitter battery – Turnigy Safety Protected 11.1v 3s 2200mAh 1.5C, $15 hobbyking.com

Motor – Rctimer 5010-14 360KV $74 for four or tiger MN3508 380kv 85gr $69ea

(Each motor thrust; 1555 prop: 3S = 700g, 4S = 950g, 5S = 1200g, 6S = 1450g) Could get more thrust with 16-17".

Speed Control –Tmotor 35A 600Hz (4) $107

Motor Batteries – Zippy 4S 5000mah battery $32

Connectors – XT60, set of 5 pairs, $3 hobbyking.com
Charger – B6AC, $35
Ground Station 12v Battery – 3S 5000mAh, B Grade $8, hobbyking.com
Extra Junk you may or may not need
    Heat Shrink Tube – 4mm, 1 foot, $0.21 hobbyking.com
    Battery straps (4), $4 hobbyking.com
    Kapton Tape, $5 eBay
    Hot glue, $1 "hair extension" high temp hot glue polyamide.
    Blue Loctite, $2 auto parts store
    2-3mm Fiberglass rods
    Fiberglass or carbon arrow shafts

Additional Options for experienced flyers

HD Camera – Nex5 style, or DSLR in octo config.
HD Camera microSD card – PNY 32GB
Gimbal- Tarot gimbal fits nice with their Tarot "card" system, also Align G800 gimbal, Cinestar, Porta Head
VTX Antenna – 1.3 GHz Skew Planar RHCP, SierraRC.com
VRX Antenna – 1.3 GHz Helical or Crosshair RHCP, SierraRC.com
RCRX Antenna – dipole, diy
RCTX Antenna – $50 arrowantennas.com 440-7ii
Notch Filter for LRS - IBCrazy 433/1.3, $20 readymaderc.com
VRX Diversity Controller + additional antenna -
Telemetry booster - 2 Watt amplifier/booster, shireeninc.com/900-mhz-2-watts-oem-module-amplifier/
    Alt: RFD900 1 watt radios
Alt: OpenLRS 2 way 1W
GPS Cellular Locator - Garmin GTU 10 Tracking Unit.
Battery upgrade –
Ground Control Case –
Video Splitter -
PC USB analog video input or DVR -
USB RTL SDR – **R820T**, $20.
Laptop to run mapping.
Tripod – $30
FPV Headset – Fatshark?
Extra Batteries –
Extra Chargers -
Extra Props –

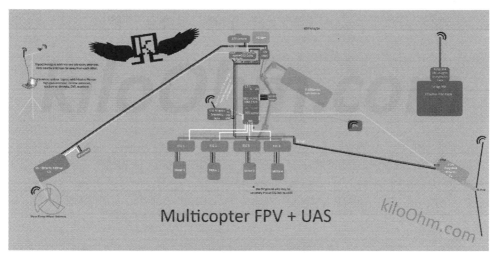

APM Multicopter FPV setup
Large image link (http://kiloohm.com/?p=483)

Acronym Definitions:

1S, 2S, 4S…etc. – Sizes of LiPo RC style batteries. The number is indicative to quantity of cells.

AC – Alternating Current. Form of electric power that is in the walls of your house, it's in your kitchen, it's in your charger, its everywhere!!!...except in your drone.

A – Amps or Amperes. Unit of electric current. See "mAh" and "Ah" definitions.

Ah – Ampere-Hour. Unit for electric charge over 1 hour. Relation to battery capacity. Example: "A motor requires 10 amps, so that 10,000 mAh (10Ah) battery will end up supplying one hour".

AI – Artificial Intelligence. Could be used in autonomous flying one day, involves the machine making decisions. Not gonna happen anytime soon. Relax, Skynet is not active....*yet*.

ARF – Almost Ready to Fly. Kit planes regularly advertise this. "Almost" is subjective. Plan on adding plenty of $ and time.

AUW – All up weight. The whole flight weight including batteries and cameras.

BEC – Battery Elimination Circuit. A simple down switching transformer to make big volts into little volts. Switching types are noisy (but tiny), so filtering with a toroid or L-C filter may be needed for FPV.

C – also "**C – Rating**". Discharge rating for batteries. Higher C ratings are better.

CAD – Computer Aided Design. Requires a dude or dude'et in a chair with a pot of coffee.

CAP – Capacitor. Like a battery kinda, but with more zap at once. Electrostatic in nature, not chemical like a battery.

COTS – Commercial Off The Shelf. Stuff you can buy in stores.

CCD – Charge Coupled Device, image sensor – camera.

CCTV – Closed Circuit Television. Smile.

CA – Adhesive, 1-10 second quickset glue. Good for pranks too! *Don't.*

CF – Carbon Fiber, fabric. Worth its weight in gold, thankfully it's lightweight.

CG – Center of Gravity. Teeter-totter at equilibrium.

CMOS – Complementary metal–oxide–semiconductor, image sensor – camera. CCTV cams.

dB – In general terms a measure of radio power. dBi is for antennas and dBm is for transmitters.

DC – Direct Current

EIRP – Total transmission power including antenna focusing.

EM – Electro-Magnetic (field)

EMP – Electro-Magnetic Pulse. OMG.

EPO – Expanded Polyolefin foam. Stiff, durable.

EPS – Expanded Polystyrene foam, coffee cups, surfboard foam. Blue or pink are *Extruded (XPS)* versions available at home improvement stores. Brittle, lightweight.

EPP – Expanded Polypropylene foam. Commonly used in electronics packaging. Durable, flexible.

FCC – Federal Communications Commission

FPV – First Person View. Inexpensive and non-resource intensive SD. Live video feed RC airplane.

GHz – 1 Gigahertz =10^9 Hz. A unit of frequency measurement. WiFi is on the 2.4GHz frequency.

GPS – Global Positioning System. Optional onboard hardware that feeds back to a laptop ground station.

g/W – Grams per Watt. A motor can carry ___ grams with ____ prop using ____ watts. AKA efficiency.

HAM – Licensed Amateur Radio Operator

HD – High Definition. Digital video of 720p or 1080p or greater resolution.

HUD – Heads Up Display. Video overlay of data: speed, altitude and other telemetry. See "OSD" definition.

Kv – RC motors use Kv to mean Rotations per Minute. Can be easily confused with KV.

KV – Kilovolts. 1KV = 1000 Volts. See "Volt" definition.

LHCP – Left Hand Circular Polarized, antenna. Not common, use RHCP.

LiPo – Lithium Polymer, battery chemistry.

mA – Milli Amp. Unit of electric current. 1000 mA = 1A. See "A" definition.

mAh – Milli Ampere-Hour. Unit for electric charge over 1 hour. Relation to battery capacity. Subunit of Ampere Hours (Ah). 1000 mAh = 1Ah

MHz – 1 Megahertz=10^6 Hz. A unit of frequency measurement. Example: Baby monitors are on the 900MHz frequency.

mW – Milliwatts. See Watt, wut.

nm – Nano-meter, unit of measure. Example: describing electro-optical wavelength; 900nm laser is in the infrared range of the electromagnetic spectrum.

Ohm – Measure of electrical impedance.

OSD – On Screen Display. Onboard hardware for visually overlaying telemetry on your video feed, AKA "HUD".

RC (R/C) – Radio Control

RCrx – Radio Control Receiver (on the plane).

RCtx – Radio Control Transmitter (in your hands or base station).

RF – Radio Frequency, radio "waves".

RHCP – Right Hand Circular Polarized, antenna.

RSSI – Received Signal Strength Indicator

RTF – Ready To Fly, yay!…but are usually junky toy planes.

Rx – Receive(r) (radio).

SD – "Standard" Definition, interlaced analog video. Fine for FPV, but not for YouTube.

SD Card – Memory Card, for storing video or picture files.

SDHC – High Capacity memory card.

SDR – Software Defined Radio. A broad spectrum radio that displays raw data on your computer screen. RTLSDR FTW!...oh FTW means For The Win…derp. I won't explain derp.

Tx – transmit(r) (radio).

UAS – Unmanned Aircraft Systems. Newspeak for UAV. Encompasses all systems such as ground control.

UAV – Unmanned Aerial Vehicle. Drone. A passé term? See "UAS" definition.

V – also "**Volt**", – A unit of measure of AC or DC electrical potential, often described as VAC or VDC or V AC or V DC.

VAC – Volts Alternating Current, V AC. See "AC" definition above. Volts, a unit of measure of electrical potential.

VDC – Volts Direct Current, V DC. See "DC" definition above. Volts, a unit of measure of electrical potential.
VRx – Video receiver (on your base station).

VTx – Video transmitter (on your plane).

W – also "**Watt**". – A unit of power measurement. Ex: Measuring radio power output.

WiFi – Wireless internet/network.

XPS – Extruded Polystyrene foam. See "EPS" definition above.

Information Resources:

Some sites and forums that are good for small RC based craft:

kiloOhm.com

diydrones.com

fpvlab.com

rcgroups.com

fpv-forum.com

I'm just one guy here and would love your feedback. If you liked this book, please give it a 5 star review! Thanks!

Made in the USA
Lexington, KY
16 December 2014